BIODIESEL FUELS REEXAMINED

ENERGY SCIENCE, ENGINEERING AND TECHNOLOGY

Additional books in this series can be found on Nova's website
under the Series tab.

Additional E-books in this series can be found on Nova's website
under the E-books tab.

ENERGY SCIENCE, ENGINEERING AND TECHNOLOGY

BIODIESEL FUELS REEXAMINED

BRYCE A. KOHLER
EDITOR

Nova Science Publishers, Inc.
New York

For permission to use material from this book please contact us:
Telephone 631-231-7269; Fax 631-231-8175
Web Site: http://www.novapublishers.com

NOTICE TO THE READER

The Publisher has taken reasonable care in the preparation of this book, but makes no expressed or implied warranty of any kind and assumes no responsibility for any errors or omissions. No liability is assumed for incidental or consequential damages in connection with or arising out of information contained in this book. The Publisher shall not be liable for any special, consequential, or exemplary damages resulting, in whole or in part, from the readers' use of, or reliance upon, this material. Any parts of this book based on government reports are so indicated and copyright is claimed for those parts to the extent applicable to compilations of such works.

Independent verification should be sought for any data, advice or recommendations contained in this book. In addition, no responsibility is assumed by the publisher for any injury and/or damage to persons or property arising from any methods, products, instructions, ideas or otherwise contained in this publication.

This publication is designed to provide accurate and authoritative information with regard to the subject matter covered herein. It is sold with the clear understanding that the Publisher is not engaged in rendering legal or any other professional services. If legal or any other expert assistance is required, the services of a competent person should be sought. FROM A DECLARATION OF PARTICIPANTS JOINTLY ADOPTED BY A COMMITTEE OF THE AMERICAN BAR ASSOCIATION AND A COMMITTEE OF PUBLISHERS.

Additional color graphics may be available in the e-book version of this book.

Library of Congress Cataloging-in-Publication Data

Biodiesel fuels reexamined / editor, Bryce A. Kohle.
    p. cm.
  Includes index.
  ISBN 978-1-60876-140-1 (hbk.)
  1. Biodiesel fuels--Environmental aspects. 2. Diesel motor exhaust gas--United States. 3. Biodiesel fuels--United States. I. Kohler, Bryce A.
  TP359.B46B5597 2011
  665'.37--dc23

                                                                                    2011033967

Published by Nova Science Publishers, Inc. ✛ *New York*

CONTENTS

PREFACE

Biodiesel, produced from seed oils or animal fats via the transesterification process, has been the focus of biofuel production because of its potential environmental benefits and because it is made from renewable biomass resources. Biodiesel can be blended with conventional diesel fuel in any proportion and used in diesel engines without significant engine modifications. In recent years, the sales volume for biodiesel in the U.S. has increased dramatically from about 2 million gallons in 2000, to 75 million gallons in 2005, to 250 million gallons in 2006. This book explores the effects of biodiesel on diesel engine nitrogen oxide and other regulated emissions.

Chapter 1- The Environmental Security Technology Certification Program (ESTCP) funded a three-year project to obtain air pollution emission factors for commonly used Department of Defense (DoD) diesel engines fueled with various types and blends of biodiesel. Biodiesel is a nontoxic, biodegradable fuel made from organic fats and oils that serves as a replacement, substitute, and enhancer for petroleum diesel. It may be blended with petroleum diesel in all existing diesel engines with little or no modification to the engines. Previous studies suggest that use of biodiesel can significantly reduce the quantity and toxicity of the air pollution produced by diesel engines.

Chapter 2- Biodiesel is a fuel-blending component produced from vegetable oils, animal fats, or waste grease by reaction with methanol or ethanol to produce methyl or ethyl esters. Pure biodiesel contains approximately 10 weight percent oxygen. It is typically blended with petroleum diesel at levels up to 20% (B20). The presence of oxygen in the fuel leads to a reduction in emissions of hydrocarbons (HC) and toxic compounds, carbon monoxide (CO), and particulate matter (PM) when biodiesel blends are burned in diesel engines. These reductions are robust and have been observed in numerous engine and vehicle testing studies. Engine dynamometer studies reviewed in a 2002 report from EPA show a 2% increase in oxides of nitrogen (NO_x) emissions for B20. This perceived small increase in NO_x is leading some state regulatory agencies to consider banning the use of biodiesel. Therefore, the issue of NO_x emissions is potentially a significant barrier to expansion of biodiesel markets.

Chapter 3- There has long been a desire to find alternative liquid fuel replacements for petroleum-based transportation fuels. Biodiesel, produced from seed oils or animal fats via the transesterification process, has been the focus of biofuel production because of its potential environmental benefits and because it is made from renewable biomass resources. Biodiesel can be derived from various biological sources such as seed oils (e.g., soybeans, rapeseeds, sunflower seeds, palm oil, jatropha seeds, waste cooking oil) and animal fats. In the United States, a majority of biodiesel is produced from soybean oil. In Europe (especially in Germany), biodiesel is produced primarily from rapeseeds. Biodiesel can be blended with conventional diesel fuel in any proportion and used in diesel engines without significant engine modifications (Keller et al. 2007). In recent years, the sales volume for biodiesel in the United States has increased dramatically: from about 2 million gallons in 2000, to 75 million gallons in 2005, to 250 million gallons in 2006 (National Biodiesel Board 2007).

Chapter 4- In support of the U.S. Department of Energy Fuels Technologies Program Multiyear Program Plan goal of identifying fuels that can displace 5% of petroleum diesel by 2010, the National Renewable Energy Laboratory (NREL), in collaboration with the National Biodiesel Board (NBB) and with subcontractor Southwest Research Institute, performed a study of biodiesel oxidation stability. The objective of this work was to develop a database that supports specific proposals for a stability test and specification for biodiesel and biodiesel blends. B100 samples from 19 biodiesel producers were obtained in December of 2005 and January of 2006 and tested for stability. Eight of these samples were then selected for additional study, including long-term storage tests and blending at 5% and 20% with a number of ultra-low sulfur diesel (ULSD) fuels. These blends were also tested for stability. The study used accelerated tests as well as tests that were intended to simulate three real-world aging scenarios: (1) storage and handling, (2) vehicle fuel tank, and (3) high-temperature engine fuel system. Several tests were also performed with two commercial antioxidant additives to determine whether these additives improve stability. This report documents completion of NREL's Fiscal Year 2007 Annual Operating Plan Milestone 10.1.

Chapter 5- Concerns about U.S. reliance on imported petroleum and fluctuating fuel prices have led to growing interest in using biodiesel, an alternative fuel made from vegetable oils. However, there is also interest in the direct use of vegetable oils as straight or raw vegetable oil (SVO or RVO), or of waste oils from cooking and other processes. These options are appealing because SVO and RVO can be obtained from U.S. agricultural or industrial sources without intermediate processing. However, SVO is not the same as biodiesel, and is generally not considered to be an acceptable vehicle fuel for large-scale or long-term use.

While straight vegetable oil or mixtures of SVO and diesel fuel have been used by some over the years, research has shown that SVO has technical issues that pose barriers to widespread acceptance.

In: Biodiesel Fuels Reexamined
Editor: Bryce A. Kohler

Chapter 1

EFFECT OF BIODIESEL ON DIESEL ENGINE NITROGEN OXIDE AND OTHER REGULATED EMISSIONS PROJECT NO. WP-0308[*]

*Bruce Holden, Jason Jack,
Wayne Miller and Tom Durbin*

EXECUTIVE SUMMARY

The Environmental Security Technology Certification Program (ESTCP) funded a three-year project to obtain air pollution emission factors for commonly used Department of Defense (DoD) diesel engines fueled with various types and blends of biodiesel. Biodiesel is a nontoxic, biodegradable fuel made from organic fats and oils that serves as a replacement, substitute, and enhancer for petroleum diesel. It may be blended with petroleum diesel in all existing diesel engines with little or no modification to the engines. Previous studies suggest that use of biodiesel can significantly reduce the quantity and toxicity of the air pollution produced by diesel engines.

The project included the measurement of the regulated air emissions of carbon monoxide (CO), hydrocarbons (HC), nitrogen oxides (NOx), and particulate matter (PM). Testing was performed in accordance with Environmental Protection Agency (EPA) testing standards and duty cycles. The tests were performed both in the laboratory and the field. The project also included measurements of Hazardous Air Pollutants (HAPs), the evaluation of two proposed NO_x reduction fuel additives, as well as the chemical speciation of the HC emissions and characterization of the PM emissions. For the project, five fuels were tested, a soy-based biodiesel, a baseline petroleum based ultra low sulfur diesel (ULSD), JP-8, and two yellow grease based biodiesels (YGA & YGB). The biodiesel fuels were tested at the 20% (B20),

---

[*] This is an edited, reformatted and augmented version of a Naval Facilities Engineering Service Center's publication, dated May 2006.

50% (B50), 70% (B70) and 100% (B100) concentration levels, with the biodiesel being mixed with the ULSD.

Ten types of DoD operated diesel engines were included in the test, including engines used for on-road, off-road, and portable power applications. Test engines were supplied by a multitude of DoD facilities. Engines were selected for inclusion in the demonstration based on their widespread use within DoD.

The primary justification for this project was to provide the biodiesel emissions data necessary to promote its increased use within DoD. Currently, there is a little emissions data in the technical literature from diesel engines of the age and types commonly used by DoD. An additional concern is the lack of data for yellow grease based biodiesel; a product manufactured using vegetable oil recycled from commercial cooking operations. It is expected that the data from this study may be incorporated with previous datasets, to provide the EPA a more detailed and comprehensive database on different varieties of biodiesel feedstocks and applications.

This project focused on B20 biodiesel blends, since this is the blend of biodiesel used in military vehicles. The project results for the regulated emissions were that at the B20 level, there were no consistent trends over all applications tested. Within the context of the test matrix, no differences were found between the different YGA, YGB, and soy-based biodiesel feedstocks. The results of more extensive statistical analyses also indicated no statistically significant differences in CO, HC, NO_x and PM emissions between the B20-YGA and the ULSD. The tested NOx reduction additives also proved to be ineffective. Thus the air pollution performance objectives outlined in the project's demonstration plan were not met. Although these results were not expected, they are not necessarily a disappointment since the baseline USLD fuel proved to be greatly superior to existing on-road Diesel No. 2.

The higher biodiesel blends (B50 to B100) were only tested on one Humvee, and with B100 on a 250 kW portable generator and a single test on the Ford F9000 tractor. On the Humvee, the higher biodiesel blends did show a trend of higher CO and HC emissions and lower PM emissions.

For the unregulated HAP emissions, no consistent trends were identified over the subset of vehicles tested. This result like those for the regulated emissions, did not meet the air pollution performance objectives outlined in the project's demonstration plan. However, it should be noted that the dataset for HAPs was smaller than that for the regulated emissions. Also, since speciation data was not available for all modes, analyses and comparisons based on weighted values could not be conducted for the species. As such, it is likely that a larger sample set would be needed to statistically evaluate the effects of biodiesel on HAPs against ULSD. While these results were not expected, it is not necessarily a disappointment since, as previously stated, the baseline ULSD fuel proved to be greatly superior to existing on-road Diesel No. 2.

Although our testing was not able to identify statistically significant air pollution benefits for the use of B20 biodiesel, from a lifecycle cost standpoint, the use of B20 is the most cost effective method for DoD fleets to meet their alternative vehicle requirements. Using B20 in place of petroleum diesel involves no new infrastructure requirements nor additional environmental compliance costs. The only cost is the $0.14 higher cost per gallon to purchase the fuel.

LIST OF ACRONYMS

AFB	Air Force Base
AFV	Alternate Fueled Vehicle
APG	Aberdeen Proving Ground
API	American Petroleum Institute
AO/AQIRP	Auto/Oil Air Quality Improvement Research Program
ARB	Air Resources Board (California)
ASTM	American Society of Testing & Materials
ATC	Aberdeen Test Center
B20	20% biodiesel by volume, 80% petroleum diesel by volume
B50	50% biodiesel by volume, 50% petroleum diesel by volume
B70	70% biodiesel by volume, 50% petroleum diesel by volume
B100	100% biodiesel by volume
BTU/GAL	British Thermal Units per Gallon
CARB	California Air Resources Board
CAT	Caterpillar Corporation
CBC	Construction Battalion Center
CBD	Central Business District
CBO	Congressional Budget Office
CCR	California Code of Regulations
CE-CERT	Bourns College of Engineering – Center for Environmental Research and Technology
CFR	Code of Federal Regulations
CH4	Methane
CO2	Carbon Dioxide
CO	Carbon Monoxide
CVS	Constant Volume Sampling
DENIX	Defense Environmental Network & Information Exchange
DLA	Defense Logistics Agency
DNPH	Dinitrophenylhydrazine
DoD	Department of Defense
DTBP	Ditertiary Butyl Peroxide
EC	Elemental Carbon
ECAM	Environmental Cost Analysis Methodology
EHN	Ethyl Hexyl Nitrate
EPA	Environmental Protection Agency
ESTCP	Environmental Security Technology Certification Program
FTP	Federal Test Procedure
G/bhp-hr	Gram Per Brake Horsepower Hour
G/mile	Gram per mile
GC	Gas Chromatography
GC-FID	Flame Ionization Detector
GC-MS	Gas Chromatography Mass Spectroscopy
GM	General Motors Corporation

HAP	Hazardous Air Pollutant
HC	Hydrocarbon
HMMWV	Humvee
HP	Horsepower
HP LC/UV	High Performance Liquid Chromatography/Ultraviolet
ISO	International Standards Organization
JP-8	Jet Propellant No. 8
KOH	Potassium Hydroxide
KW	Kilowatt
LLC	Limited Liability Corporation
MEL	Mobile Emissions Laboratory
MSDS	Material Safety Data Sheet
MSAT	Mobile Source Air Toxics
NAAQS	National Ambient Air Quality Standards
NDIR	Non Dispersive Infrared
NFESC	Naval Facilities Engineering Service Center
NMHC	Non Methane Hydrocarbon
NIOSH	National Institute of Occupational Safety and Health
NOx	Nitrogen Oxides Chemical Compounds Including NO and NO2
NREL	National Renewable Energy Laboratory
NYBC	New York Bus Cycle
OC	Organic Carbon
OSHA	Occupational Safety and Health Administration
PAH	Polycyclic Aromatic Hydrocarbon
PM	Particulate Matter
PPM	Parts Per Million
POC	Point of Contact
PSI	Pounds per square inch
QA	Quality Assurance
QC	Quality Control
ROVER	Real-time On-road Vehicle Emissions Reporter
SAE	Society of Automotive Engineers
SO2	Sulfur Dioxide
TC	Total Carbon
THC	Total Hydrocarbon
UCR	University of California, Riverside
ULSD	Ultra Low Sulfur Diesel
US06	EPA Aggressive Certification Cycle
VOC	Volatile Organic Compound
YGA	Yellow Grease Formula A
YGB	Yellow Grease Formula B

ACKNOWLEDGMENTS

The Project Team would like to thank the dedicated employees at the Camp Pendleton Marine Corps Base for making the arrangements to supply most of the diesel engines tested in this project. Camp Pendleton was able to supply the engines at no cost to the project and accommodate numerous delays in returning the engines. The team would also like to thank Biodiesel Industries, Inc. and the National Renewable Energy Laboratory for supplying 400 gallons of yellow grease biodiesel and emissions testing services respectfully, at no cost to the project. Finally, the Project Team would like to thank the Environmental Security Technology Certification Program office for providing the majority of the project funding. .

Contact information for the main project participants is provided in Paragraph 8. The ESTCP Cost and Performance Report may be found at their WEB site: http://www.estcp.org.

1. INTRODUCTION

1.1. Background

Diesel engines are widely used throughout the DoD for powering tactical and non-tactical vehicles and vessels, off-road equipment, engine-generator sets, aircraft ground-support equipment and a variety of other applications. Like gasoline engines, diesels are known to emit all of the criteria pollutants regulated by the National Ambient Air Quality Standards (NAAQS) established by the Clean Air Act. Human health concerns with diesel exhaust are; however, primarily focused on PM and HAP emissions.

Although diesels are the most efficient of internal combustion engines and have favorable characteristics in the reduction of green-house gas emissions, concerns with the health effects from PM and HAP emissions has intensified the call for cleaner burning diesels and lead to recently proposed and enacted regulations increasing restrictions on diesel exhaust emissions. Because of these developments, many control approaches are being pursued. One solution is the development of cost-effective alternative fuels, such as biodiesel, to reduce diesel engine emissions.

Biodiesel is a nontoxic, biodegradable fuel made from organic fats and oils, and serves as a replacement, substitute and enhancer for petroleum diesel. It may be used in all existing diesel engines with little or no modification to the engines. Biodiesel has been previously reported (see Reference 7.1) to reduce all regulated air pollutant emissions except for emissions of NO_x. Biodiesel may be blended with petroleum diesel at any percentage For DoD applications, it is customary to use a 20 percent by volume biodiesel 80 percent by volume petroleum diesel (B20), [pure biodiesel = (B100)] biodiesel blend. Other major biodiesel consumers commonly use B2, B5, B11 or B20 biodiesel blends, as well as neat biodiesel (B100).

Alternative fuels are mandated for all federal fleets of 20 or more vehicles and B20 is one of the options for meeting this requirement. Biodiesel has been designated as an Alternative Fuel by the Department of Energy, and has been registered with the EPA as a fuel and fuel additive. Authorization for biodiesel use by DoD in non-tactical vehicles was approved in 1999.

Although there is much support for the continued development of the biodiesel alternative, there is not currently sufficient knowledge on how various types and blends of biodiesel affect the air emissions from diesel engines of interest to DoD. Specifically, there are little data on the emission benefits of biodiesel produced from used vegetable oils. These data are important since significant quantities of used vegetable oils for the production of biodiesel are currently available near and on DoD facilities at little to no cost. It has been reported to the author that many DoD facilities are in fact, paying for the disposal of their used vegetable oil.

1.2. Objectives of the Demonstration

The objective of this project is to establish emissions factors for DoD diesel powered engines of interest fueled with various types of biodiesel. Currently, most of the available biodiesel emissions data is for older heavy-duty engines tested on an engine dynamometer and fueled with virgin soybean derived biodiesel. Although these data are important, their use in estimating DoD fleet emission factors introduces significant uncertainties. Previous Naval Facilities Engineering Service Center (NFESC) surveys have shown that although most DoD non-tactical diesel powered engines are heavy duty, a high percentage are newer engines that employ emission control technologies that are significantly different than the tested engines. In addition, it is expected that a significant portion of DoD biodiesel programs in the future will employ yellow grease derived biodiesel made from recycled vegetable oil. By targeting the emissions from actual DoD operated heavy-duty engines fueled with either soybean or yellow grease derived biodiesel, relevant DoD emissions factors can be determined.

A secondary objective of this test program is to identify and demonstrate fuel additives that reduce NO_x emissions from biodiesel. While biodiesel has been shown to reduce air emissions of the other criteria pollutants, numerous studies have shown that its use results in a slight increase in NOx emissions (i.e., < 2 percent for B20). Research at Arizona State University suggests that the addition of a cetane improver such as ethyl hexyl nitrate (EHN, 1/2 percent by volume) or ditertiary butyl peroxide (DTBP, 1 percent by volume) will reduce NOx emissions from biodiesel.

Another factor increasing the importance of this emissions testing program is the total lack of data comparing the emissions from ULSD with those from biodiesel. Starting in 2006, the EPA has mandated the use of diesel fuel with a sulfur level < 15 ppm for on-highway applications. A similar requirement has been proposed for off-highway use. For this test program, NFESC will exclusively compare various types of biodiesel with ULSD for the testing of non-tactical vehicles since this test program will complete within one year of the ULSD rollout.

In addition to the measurements of currently regulated emissions, the test program will also include PM chemical analysis as well as the emissions of a number of hazardous air pollutants. This work will focus on yellow grease where the least information is available.

The testing program will include eight types of DoD operated vehicles and two portable engines. Not all the same test cycles nor fuels will be used for each test engine. Multiple testing locations with different capabilities will be used. Test results will be reported in a common format to simplify comparisons made between all the test runs.

For this project, NFESC will obtain the biodiesel emissions data necessary for DoD decision makers to intelligently plan future biodiesel implementations. Biodiesel air emissions data will be provided to the Defense Logistics Agency (DLA) along with other organizations that control specifications for DoD fuel purchases. This project will address Navy need 2.I.01.b, "Control Particulate and Other Air Emissions from Mobile and Stationary Sources", and Air Force need 506, "Eliminate NOx Emissions from Fuel-burning".

1.3. Regulatory Drivers

Mobile-source diesel emissions are regulated by both Federal (40 CFR 86, 89) and California (13 CCR Chapter 3) equipment and vehicle standards. These standards are applied to equipment and vehicles at the time of manufacture. In the last six years, EPA has pursued a program to dramatically tighten these regulations. This is illustrated in Table 1.1 below, which shows the 2007 EPA on-road heavy-duty engine standards, along with the year 2000 and 2004 standards. Likewise, the EPA has also pursued a program to dramatically tighten the regulations for non-road diesel engines. These regulations, unlike their on-road counterparts, are based on the size of the engine, with larger engines having tighter standards.

Table 1.1. Current and Future EPA Emissions Regulations [g/bhp-hr]

		2000 Standard (g/bhp –hr)	2004 Standard (g/bhp –hr)	2007 Standard (g/bhp –hr)	Phase-In by Model Year*			
					2007	2008	2009	2010
Diesel Fleet	NO_x	4.0	N/A	0.20	25%	50%	75%	100%
	HC	1.3	N/A	0.14				
	NMHC + NO_x	N/A	2.4	N/A				
	CO	15.5	15.5	15.5	100%	100%	100%	100%
	PM	0.10	0.10	0.01	100%	100%	100%	100%

* Percentages represent percent of sales

The 2007 heavy-duty highway diesel engine standards will reduce PM emissions by about 98 percent from a 1990 baseline and 90 percent from a 2000 baseline. Significant NO_x and non-methane hydrocarbon (NMHC) reductions, are also required for 2004 and later engines. However, because these emission decreases do not affect existing diesel engines, their full benefit will take more than 20 years to achieve. In an effort to achieve the benefits sooner, several states have proposed regulatory strategies to reduce emissions for existing engines.

In October 2000, the California Air Resources Board (CARB) finalized their *Risk Reduction Plan to Reduce Particulate Matter Emissions from Diesel-Fueled Engines and Vehicles.* The California plan calls for the use of low-sulfur fuels, retrofit requirements, or the replacement of existing engines for on-road, non-road, portable, and stationary equipment.

In 2001, Texas enacted regulatory changes to reduce emissions from diesel engines. Their plan is a comprehensive set of incentive programs. The plan includes: 1) The Retrofit and Repower Incentive Program for On-Road and Non-Road High-Emitting Engines, 2) The New

Purchase and Lease Incentive Programs for Light-Duty and Heavy-Duty On-Road Vehicles, and 3) Clean diesel fuel requirements which include limitations on aromatics and sulfur in commercial diesel fuels. All of these changes will reduce both NO_x and particulate emissions.

Stationary-source diesel emissions are regulated by state and local regulations. Currently, most regulations only limit CO, NO_x, and opacity. However, CARB recently proposed guidance that if adopted by local air districts would require the reduction of HAP emissions.

In addition to the air emissions regulations, federal policymakers have also established several initiatives that require the use of alternative transportation fuels such as biodiesel. The purpose of these initiatives is to reduce the nation's oil imports. The Federal Fleet Acquisition Requirement in the Energy Policy Act (i.e., Title III) requires that 75 percent of annual DoD light duty vehicle acquisitions be capable of operating on alternative fuels. This law pertains to federal vehicle fleets consisting of 20 or more centrally fueled vehicles. Executive Order 13149, *"Greening the Government through Federal Fleet and Transportation Efficiency,"* requires that federal fleets reduce petroleum consumption by 20 percent by 2005, compared with the 1999 levels.

1.4. Stakeholder/End-User Issues

As described in paragraph 1.3, DoD fleet operators are under increasing pressure to reduce both diesel air emissions and petroleum consumption. Unfortunately, many alternate fuels that have been shown to reduce emissions either have fuel costs higher than petroleum diesel or require significant engine modifications and/or infrastructure upgrades. Ideally, an alternate fuels program must be cost effective, universally applicable, and provide significant measurable environmental benefits. As demonstrated by NFESC at their Port Hueneme, California research biodiesel production facility, only biodiesel derived from yellow grease meets these requirements. It is cost effective, it can be used without any engine modifications and it does not require any infrastructure upgrades. Unfortunately, at this time, it is not available throughout the country. Also, since its raw material is limited, its supply cannot meet all of the potential DoD demand. Virgin soybean derived biodiesel, an approved alternate fuel, is more widely available and can supply all of the potential DoD demand. Unfortunately, it is more expensive to produce than petroleum based diesel. This cost difference is, however, almost completely made up by existing federal subsidies. This subsidy is not permanent and therefore could be reduced or eliminated at any time in the future.

In the last couple of years, a significant number of DoD fleets have already made the switch to B20 biodiesel. The Air Force, Marine Corps and Navy have, in fact, switched most of their non-tactical vehicles. At this time, many additional DoD fleet operators are also considering switching to B20. Many of these potential DoD B20 customers are concerned about the cost of the fuel and its effect on air pollution regulatory compliance. By providing emissions testing results for multiple types of biodiesel, this project should address these potential customer concerns.

2. TECHNOLOGY DESCRIPTION

2.1. Technology Development and Application

Biodiesel is a renewable, clean burning, oxygenated fuel for diesel powered engines or boilers made from soybean or other vegetable oils or animal fats. Chemically, biodiesel consists of a small number of alkyl esters. It contains no sulfur or aromatics and already meets the EPA's 2006 on-road standard for sulfur content in diesel fuel. Because it has properties similar to petroleum-based diesel fuel, biodiesel can be blended in any ratio with petroleum diesel and used in diesel engines without major modifications. Biodiesel is registered as a fuel and fuel additive with the EPA and meets clean diesel standards established by CARB.

Biodiesel use as a fuel is as old as the diesel engine. Rudolph Diesel, inventor of the diesel engine in 1892, used peanut oil as the original engine fuel. The use of petroleum fuels for diesel engines only came into widespread use in the 1920's. This fuel substitution was the result of a significant drop in the price of petroleum. Starting in the 1970's after the "oil crisis", interest in the use of domestically produced biofuels returned. In the 1990's, the biodiesel industry organized to promote its use. Recently, biodiesel demand has mostly come from fleet operators affected by the 1998 EPAct Amendment.

Two recent successes have helped advance the widespread use of biodiesel. First, the American Society of Testing and Materials (ASTM) issued a specification (D 6751) for biodiesel fuel in December 2001. ASTM is the premier standard-setting organization for fuels and additives in the United States. This development is crucial in standardizing fuel quality for biodiesel in the U.S. market and increasing the confidence of consumers and engine makers. The ASTM specification was developed so that approved fuels could be consistently manufactured using any vegetable oil or animal fat as the raw material. Second, biodiesel became the only alternative fuel in the country to have successfully completed the EPA's Tier I and Tier II Health Effects testing under Section 211(b) of the Clean Air Act in May 2000. The Tier I testing conclusively demonstrated biodiesel's significant reductions in most currently regulated emissions as well as most unregulated emissions—especially those associated with cancer and lung disease. Tier II testing demonstrated biodiesel's non-toxic effect on health.

The production of biodiesel is based on the process of base-catalyzed transesterification at low temperature (150 °F), low pressure (20 psi), and with a high conversion factor (98 percent). As depicted in Figure 2.1, a fat or oil is reacted with an alcohol (like methanol) in the presence of a catalyst to produce glycerine and methyl esters or biodiesel. The methanol is charged in excess to assist in quick conversion and unconverted methanol is recycled. The catalyst is usually sodium or potassium hydroxide that has already been mixed with the methanol.

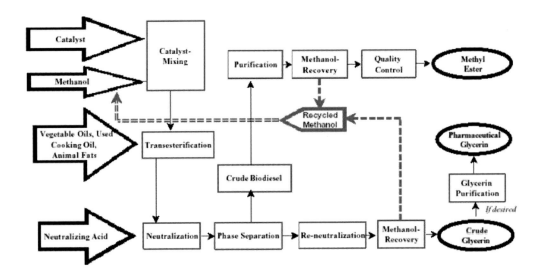

Figure 2.1. Biodiesel production diagram.

The key chemical reactions are shown below:

$$
\begin{array}{lllll}
CH_2OCOR''' & & & CH_2OH & R'''COOR \\
| & & \text{Catalyst} & | & \\
CH_2OCOR'' & + \quad 3\ ROH & \text{------>} & CH_2OH & + \quad R''COOR \\
| & & & | & \\
CH_2OCOR' & & & CH_2OH & R'COOR
\end{array}
$$

| 100 pounds | 10 pounds | 10 pounds | 100 pounds |
| Oil or Fat | Alcohol (3) | Glycerin | Biodiesel (3) |

2.2. Previous Testing of the Technology

During the past 20 years, more than 80 scientific studies have been conducted to measure the emissions from heavy-duty diesel engines fueled with biodiesel. Although the studies had many different focuses, most of the work was done using older engines (i.e., pre 1998), with testing performed on an engine dynamometer as opposed to an actual diesel powered vehicle. These studies primarily tested biodiesel derived from soybean oil since it is the most common form of the fuel.

In October 2002, EPA issued the draft technical report EPA 420-P-02-001 "A Comprehensive Analysis of Biodiesel Impacts on Exhaust Emissions". In this report (see Reference 7.1), EPA used various statistical analytical tools to compile the results from 39 studies. Wherever sufficient information was available, EPA attempted to develop models to predict how biodiesel emissions would be affected by various duty-cycle, engine age/type, and fuel properties. In addition, they summarized the results to identify the average expected emissions reductions. For use of B20, Table 2.1 provides the expected criteria pollutants emissions reductions for virgin soybean-based biodiesel added to an average low sulfur (i.e., <500 ppm) base fuel.

**Table 2.1. Emission Impacts of B20 for Soybean-Based
Biodiesel Added to an Average Base Fuel**

Regulated Pollutant	Percent Change in Emissions for Soy	
NO_x	+ 2.0%	
PM	- 10.1	%
HC	- 21.1	%
CO	- 11.0	%

The EPA analysis noted that biodiesel impacts on emissions varied depending on the type of biodiesel (i.e., manufactured from soybean, rapeseed, or animal fats) and on the type of conventional diesel to which the biodiesel was added. For example, biodiesel based on yellow grease provided a greater environmental benefit in that the reduction was greater for CO, HC, and PM and the increase in NO_x was less than with soy-based biodiesel. With one minor exception, emission impacts of biodiesel did not appear to differ by engine model year.

2.3. Factors Affecting Cost and Performance

As presented by the National Biodiesel Board, a biodiesel industrial trade organization, at their 2006 National Biodiesel Conference & Expo, for biodiesel manufactured from virgin soybean oil, the feedstock costs account for approximately 70 percent of the direct production costs, including the plant capital costs. For example, it takes about 7.5 pounds of soybean oil costing about 21 cents per pound to produce a gallon of biodiesel, thus feedstock costs alone are at least $1.58 per gallon. With processing, marketing and overhead expenses and profit included, the price of the finished biodiesel is typically over $3 per gallon.

Biodiesel producers are trying to reduce these feedstock costs by a variety of methods including developing higher oil-content soy hybrids, using other vegetable oils with a higher oil content or using yellow grease that is often available at low (~5 cents per pound) or no cost. By employing one of these strategies, it is estimated that the future cost of the feedstock can be significantly reduced, thus making biodiesel less expensive than petroleum-based diesel fuel. The Department of Energy has forecasted that biodiesel manufactured from yellow grease will cost approximately $1.40 per gallon and mustard-based biodiesel will cost less than $1 per gallon by 2010.

Two performance issues are of primary concern with the use of biodiesel. The first concern is how it affects the emissions from diesel engines. Information on emissions affects is shown in Table 2.1. The second concern is how the use of biodiesel will affect the fuel economy of a diesel engine. As reported by the EPA in their Draft Technical Report EPA 420-P-02-001 (Reference 7.1), the average virgin soybean oil derived biodiesel has an energy content of 119k Btu/gal, compared with an average 130k Btu/gal for petroleum diesel. This basic energy content difference results in a lower fuel economy for biodiesel. The EPA in their report included a summary of 217 actual fuel economy tests. Their results showed that, the fuel economy for B20 was 0.9 – 2.1 percent less than for petroleum diesel.

2.4. Advantages and Limitations of the Technology

The use of biodiesel fuel has been shown (see Reference 7.1) to reduce the overall air pollution resulting from diesel engine operations. Of the criteria pollutants regulated by the EPA, biodiesel is reported to reduce CO, HC, PM, and Sulfur Dioxide (SO_2) emissions while only causing a small increase in NO_x emissions. It also has been shown to reduce HAP and greenhouse gas emissions. In addition to its pollution reduction advantages, biodiesel also has economic and strategic advantages. Biodiesel can be made from domestically produced agricultural raw materials that are produced in surplus in the United States. The use of biodiesel will reduce the use of imported oil, much of which is supplied by potentially unstable Middle Eastern suppliers.

Although biodiesel produced from virgin raw materials such as soybean oil is more expensive than petroleum, its use has significant economic benefits. By employing the surplus of raw agricultural products, the cost of government crop support programs can be reduced. In addition, biodiesel production facilities produce employment opportunities, particularly in rural areas.

To address the raw material cost problem, the production of biodiesel manufactured from yellow grease, a food service waste product, is being greatly expanded. Currently, yellow grease is available at little to no cost. In most cases, food service operators are required to pay for the disposal of their yellow grease. Unfortunately, the supply of yellow grease is not unlimited. It is estimated that up to 800 million gallons of yellow grease may be available in the United States, a quantity sufficient to produce 700 million gallons of B100 biodiesel. This quantity cannot supply all of the potential demand. The Energy Information Administration of the Department of Energy reported in their Annual Energy Review that 2.455 million barrels per day of petroleum diesel fuel was used by the transportation sector in the United States during year 2002. If biodiesel was used for all diesel transportation needs, yellow grease could supply 9.8% of the potential B20 demand.

Three issues potentially limit the widespread and growing usage of biodiesel. The first is the cost of biodiesel and that is primarily driven by the feedstock cost. Hence in this project, we are trying to use yellow grease to reduce the fuel cost. A second potential limitation is that there has been reported a small increase in the NO_x emissions and any increase might not be acceptable in NO_x non-attainment areas. Accordingly, this project will test some additives that claim to reduce the increase in NO_x that is associated with biodiesel. The third issue is the stability and cold weather performance of biodiesel. Currently there is little information and no recognized test procedure to measure biodiesel's long-term stability. The use of acid number has however, been suggested as the best simple test method to measure biodiesel stability. Likewise, biodiesel's stability in cold weather is not well understood, although it has been shown that B100 may not be suitable in very cold weather applications, such as winter use in Minnesota. These stability concerns have currently limited DoD biodiesel usage to B20 for applications where the fuel will not be stored for extended periods.

To assess the overall potential air pollution control benefit of implementing a biodiesel program, it must be evaluated against the potential alternatives. Biodiesel is a fuel-based solution to controlling emissions from diesel engines. Alternative controls include either expensive add-on devices or replacing the engine with one that meets tougher emission standards. These emission control approaches are much more expensive than a simple fuel change; hence, biodiesel may achieve the targeted reduction in emissions at a lower total cost.

In addition to providing a low cost option to reduce diesel engine emissions, it is also the low cost option for implementing the EPAct regulations. These regulations require specified fleet operators, including most DoD fleets, to use alternate fueled vehicles (AFV) for at least 75 percent of their fleet. Using 450 gallons of B100 or 2,250 gallons of B20 earns the fleet operator one AFV credit. The Congressional Budget Office (CBO) determined in 1998 that using B20 biodiesel is the lowest cost option among the alternative fuel choices available to meet AFV requirements. The CBO predicted that the federal government would save $10 million annually by using B20 biodiesel in its fleet vehicles.

3. DEMONSTRATION PLAN

3.1. Performance Objectives

Table 3.1. Performance Objectives

Type of Performance Objective	Primary Performance Criteria	Expected Performance (Metric)	Actual Performance (future) Objective Met?
Quantitative	Reduce CO Emissions	Reduce emissions by 9% minimum with B20	No
	Reduce HC Emissions	Reduce emissions by 16% minimum with B20	No
	Reduce PM Emissions	Reduce emissions by 8% minimum with B20	No
	Reduce HAP Emissions	Reduce emissions by 16% minimum with B20	No
	Minimize Increase in NO_x Emissions	Emissions increase < 3%, for B20 (Without additive)	Yes
	Reduce NO_x Emissions	Reduce emissions by 2%, minimum for B20 (With additive)	No
	Minimize Increase in Fuel Consumption	Increase fuel consumption by 3% maximum with B20	Yes
Qualitative	Drivability	No change	Yes
	Maintain Reliability	No breakdowns caused by B20 biodiesel	Yes

Note: The performance objectives are based on a comparison with ULSD.

Table 3.2. Biodiesel Emissions Test Matrix

Item No.	Test Location	Application Description	Owner/ Operator	Engine Make/Model	Model Year	Fuel Type/ Fuel Additive	Test Cycle/Load	Regulated Emissions	HC & PM Characterization
1	NREL / UCR Mobile Lab	Thomas Bus License No. G32 001589	Cheyenne Mountain Air Station	Cummins 5.9L	2002	ULSD B20 (soy)	Cheyenne Mountain Custom Cycle	All	None
2	UCR	HMMWV (Humvee)	Camp Pendleton	GM 6.5L Model A2	2004	ULSD JP-8 B20 (YGA) B50 (YGA) B70 (YGA) B100 (YGA) B20 (soy) B100 (YGA) B100 (YGA) + Additive 1 B100 (YGA) + Additive 2	FTP, US06,	All	ULSD B20 (YGA) B100 (YGA) FTP Modes Only
	ATC		Aberdeen Proving Grounds	GM 6.2L Model A1 M998	1987	ULSD B20 (YGA)	In-use	CO, NO$_x$	None
3	ATC	Harlan Aircraft Tug	Aberdeen Proving Grounds	Cummins C6 3.9L Engine	1999	ULSD B20 (YGA)	In-use	CO, HC, NO$_x$	None
4	UCR	Stake Truck, Ford F700 Series License No. G71	Naval Base Ventura County	Cummins 5.9L – 175HP	1993	ULSD B20 (YGB) B20 (soy)	8-mode	All	None
5	UCR	Tractor, Ford L-9000 License No. MC 288060	Camp Pendleton	Caterpillar 3406C	1992	ULSD B20 (YGA) B20 (YGB) B100 (YGA) 49 State EPA No. 2 Diesel	8-mode	All	ULSD B20 (YGA) B100 (YGA)

Table 3.2. (Continued).

Item No.	Test Location	Application Description	Owner/ Operator	Engine Make/Model	Model Year	Fuel Type/ Fuel Additive	Test Cycle/Load	Regulated Emissions	HC & PM Characterization
6	ATC	Hyster 65 Forklift Model H65XM VIN No. H177B257804	ATC	Perkins 2.6L - 55HP Engine family 1PKXL02.6U B1	2001	ULSD B20 (soy)	In-use	HC, CO, NO$_x$	None
7	UCR	Ford F-350 Pickup License No. MC 291724	Camp Pendleton	Navistar 7.3 L	1999	ULSD B20 (YGA)	FTP US06	All	None
8	UCR	Thomas Bus License No. G32 00583	Camp Pendleton	CAT 3126, 330 HP	2000 Engine 1999	ULSD B20 (YGA) B20 (soy)	8-Mode	All	All
9	UCR	Portable 250 KW Generator	Camp Pendleton	Kamatzu SA60125E-2	2000	ULSD JP-8 B20 (YGA) B100 (YGA)	5-Mode	All	ULSD B20 (YGA) B100 (YGA)
10	UCR	60 KW Tactical Generator	Camp Pendleton	Lippy MEP-806A	1995	ULSD JP-8	5-Mode	All	None

Notes:
1. The acronyms for the test locations are the National Renewable Energy Laboratory (NREL) located in Denver, CO., the University of California Riverside (UCR) located in Riverside, CA, and Aberdeen Test Center (ATC) located at Aberdeen Proving Ground (APG), MD.
2. For engine no. 2, two fuel additives are listed. The purpose of both of the additives is to reduce NOx by increasing the cetane number of the fuel. Additive no.1 is ethyl hexyl nitrate (EHN, 1/2 percent by volume) and additive no. 2 is ditertiary butyl peroxide (DTBP, 1 percent by volume).
3. Two types of yellow grease were tested, YGA and YGB. These fuels were supplied from independent sources.
4. None of engines to be tested will have Catalyzed Soot Filters installed. Some of the engines may be equipped with a Diesel Oxidation Catalyst.

3.2. Selecting Test Sites/Facilities

Biodiesel emissions testing was performed at laboratory test facilities and DoD activities. The laboratory test facilities were selected based on their capabilities, proximity to NFESC, costs, and most importantly, their willingness to participate in this testing program. Field-testing sites consists of DoD facilities that operate diesel engines of interest. These sites were selected based on the availability of diesels of interest as well as their willingness to participate in the test program. The decision as to where to perform each of the emissions tests was based on many factors, including the owner's needs, the capability of the test personnel to perform field measurements, and costs.

For this emissions testing program, eight types of diesel powered vehicles and two portable engines were selected for testing. These engines were selected since they represent a good cross section of diesel engines commonly found at DoD bases as verified during previous NFESC surveys. Test engines were selected to provide the greatest possible array of equipment. They included on-highway, off-highway, military tactical, and portable power equipment. A primary consideration in the selection of the test units was the equipment operating profile and the number of units in the DoD inventory. Here emphasis was placed on equipment that normally operates at medium to high load levels, with long operating times. Information on the selected engines and the tests to be performed is provided in Table 3.2. Pictures of some of the different test engines and vehicles are provided in Appendix A.

.3. Test Site/Facilities History/Characteristics

The equipment selected for testing is located at the DoD facilities described below:

U.S. Army Aberdeen Test Center (ATC), Aberdeen Proving Grounds, MD, is a temperate-climate proving ground encompassing 57,000 acres of land and water. It is DoD's lead test center for land vehicles, guns and munitions, and live-fire vulnerability and lethality testing. After more than 80 years, ATC has developed into a world-class, all-purpose test center operating as an outdoor laboratory. The comprehensive array of capabilities, unique facilities, simulators and models at ATC, combined with an experienced scientific and technical workforce, enable testing and experimentation on items ranging from components to entire systems. To support its testing mission, many of the diesel vehicles used by the DoD are found on Aberdeen Proving Grounds.

Cheyenne Mountain Air Station, Colorado Springs, CO, is buried 2,000 feet under Cheyenne Mountain. The facility is situated in underground tunnels that were bored out of the mountain. The air station is a top-secret combat operations center formerly known as the North American Air Defense Command, or NORAD. The station contains equipment that provides warning of missile or air attacks against North America and can serve as the focal point for air defense operations in the event of an attack. The station's mission is to provide Canadian and U.S. National Command authorities with accurate air, space, missile and nuclear detonation information. The major units of the station are the North American Aerospace Defense Command, U.S. Space Command, and Air Force Space Command. To access the main operational areas, diesel powered vehicles are used in the underground tunnels. Exhaust from these vehicles is the major source of contamination for the facility's air handling system. The Thomas buses selected for the demonstration are used to transport workers down the main access tunnel.

Marine Corps Base, Camp Pendleton, CA, is the site of the Corps largest amphibious assault training facility, encompassing 17 miles of Southern California coastline and 125,000 acres. The base has a population of nearly 40,000 Marines and Sailors. As such, nearly all types of equipment in the Marine Corps inventory are located at this facility. As a functioning training command, the equipment is used almost daily for training and transportation purposes. The buses and trucks selected for testing are used to transport Marines and equipment to the widely separated training ranges within Camp Pendleton, and to other Marine Corps activities. These selected buses and trucks are also commonly found at numerous other DoD facilities.

Naval Base Ventura Country, Port Hueneme Site, Port Hueneme, CA, is the home of the Construction Battalion Center (CBC), the command organization for the Navy's "Seabees". The site covers approximately 1,600 acres on the Southern California coastline and includes a deep-water port facility. To support the Seabees in their field construction mission, the CBC has a wide variety of diesel powered vehicles and equipment much of which is extensively used during training exercises. The testing of this equipment/vehicles will be a good representation of the types of diesel engines encountered at other Navy shore activities.

Laboratory testing facilities to be used for this program are described below:

University of California, Riverside (UCR), CA, Bourns College of Engineering – Center for Environmental Research and Technology (CE-CERT) is an off-campus air emissions testing and air pollution research facility of the University of California, a state supported institute of higher education. CE-CERT was founded over 12 years ago to support California's effort to understand and reduce air pollution. As part of its capabilities, CE-CERT has two emissions testing facilities that will be used during this project. One is a light-duty chassis dynamometer emissions testing facility and the other is a mobile test laboratory contained in a trailer. Both testing facilities are capable of measuring all of the criteria pollutants on various engine-operating cycles in accordance with EPA approved test methods. In addition to providing emissions rate data, these facilities can also provide HC and PM speciation measurements as well as PM size distribution measurements.

The UCR mobile laboratory was designed to be pulled by a heavy-duty tractor. This allows real on-road emissions measurements to be obtained. The mobile laboratory's design also allows it to be used to measure emissions rates from stationary diesel sources such as back-up generators.

National Renewable Energy Laboratory ReFUEL Laboratory, Denver, CO, made its debut in 2002 to provide facilities for identifying, testing, and evaluating renewable and synthetic fuels and lubricants for use in ground transportation, with a focus on enabling high efficiency operation while displacing petroleum products. The 4,500 square foot facility was previously operated by the Colorado School of Mines as the Colorado Institute for Fuels and High Altitude Engine Research. It was designated as the National High Altitude Heavy-Duty Research and Technology Center under the Clean Air Act Amendments of 1990. The facility includes a heavy-duty chassis dynamometer with tandem 40-inch rolls capable of testing single or tandem drive axle vehicles up to 80,000 lbs, and a 24-foot wheelbase. It is the only high altitude facility of its type in North America. The facility provides air pollution measurement equipment to measure all criteria pollutants using EPA approved test methods and driving cycles. In addition, it has an engine test cell for directly testing engines.

Table 3.3. B100 Biodiesel Chemical/Physical Tests

Property*	ASTM Test Method	Limits	Units	Yellow Grease A	Yellow Grease B	Soy-Biodiesel
Acid Number	D 664	0.80 max.	mg KOH/g	0.2	0.36	0.45
API Gravity	D287			29.1	29.3	28.7
Specific Gravity				0.881	0.880	0.883
Btu Content – Net Heating Value	D240		Btu/gal	126,344	122,355	121,618
Carbon Residue	D 4530	0.050 max.	% Mass	0.01%	0.013	<0.01
Cetane Number	D 613	47 min.		52.7	54.1	54.3
Cloud Point	D 2500	Report	°C	4	4	-2
Copper Strip Corrosion	D 130	No. 3* max.		1a	1a	1a
Distillation at 90%	D 86	360 max.	°C	352	352	352
Flash Point	D 93	130.0 min.	°C	>160	200	141
Free Glycerin	D 6584	0.020	% Mass	0.000	0.012	0.004
Kinematic Viscosity, 40°C	D 445	1.9 – 6.0	mm^2/s	3.807	4.464	4.086
Phosphorous Content	D 4951	0.001 max.	% Mass	0.0003	0.0000	0.0000
Sulfated Ash	D 874	0.020 max.	% Mass	0.001	0.008	0.000
Sulfur	D 5453	0.05 max.	% Mass	0.0011	0.00324	0.00005
Total Glycerin	D 6584	0.24	% Mass	0.098	0.158	0.01
Water and Sediment	D 2709	0.050 max.	% Volume	0	0	0

Note: ASTM D6751, the specification for biodiesel fuels, requires the fuel to meet the properties identified in ***Bold and Italic.*** * Comparison to a color chart for corrosion.

3.4. Present Operations

All 11 pieces of equipment proposed for emissions testing utilize diesel engines manufactured by various manufacturers between the years 1987 and 2004. The engines are all used at DoD facilities. Many of the engines produce visible soot during operation, making them prime candidates for application of a clean fuel program.

Table 3.4. Petroleum Diesel Chemical/Physical Tests

Property	ASTM Test Method	Limits	Units	CARB Certified ULSD	JP-8
API Gravity	D287			38.5	39.3
Specific Gravity				0.832	0.828
Aromatics	D1319		% Vol.	19.3	16.0
Btu Content – Net Heating Value	D240		Btu/gal	128,413	127,530
Rams Carbon Residue on 10% btms.	D 524	0.350 max.	% Mass	0.1	0.05
Cetane Number	D613	40 min.		54.4	36.3
Cetane Index	D976			51.92	36.4
Cloud Point	D 2500	Report	°C	-5	<-21
Copper Strip Corrosion	D 130	No. 3* max.		1a	1a
Distillation at 90%	D 86	338 max.	°C	328	242
Flash Point	D 93	130.0 min.	°C	159.8	130
Kinematic Viscosity, 40°C	D 445	1.9 – 6.0	mm²/s	2.602	1.484
Ash	D 482	0.010 max.	% Mass	<0.001	<0.001
Sulfur	D 5453	0.05 max.	% Mass	0.0002	0.0461
Water and Sediment	D 2709	0.050 max.	% Volume	0	0

Note: ASTM D975, the specification for petroleum diesel fuels, requires the fuel to meet the properties identified in **Bold and Italic.** * Comparison to a color chart for corrosion.

3.5. Pre-Demonstration Testing and Analysis

Prior to initiating the actual emissions testing program, samples of all program fuels were analytically tested for specified chemical and physical properties. Fuel analysis results are provided in Table 3.3 for biodiesel fuels and 3.4 for petroleum fuels. These tests have been selected based on the ASTM specifications for the fuels and as recommended by the EPA in Reference 7.1.

3.6. Testing and Evaluation Plan

3.6.1. Demonstration Set-Up and Start-Up

The emissions testing program took place both at laboratory sites and in the field. In Table 3.2, the test location for each engine test is identified. When the testing was performed at a laboratory, the test engine was transported to the laboratory. As required, fuels were transported to the test site. Fuels and fuel additives for the testing were stored and mixed at UCR. To reduce the variability of the test results due to changes in fuel composition, all fuels required for the project were purchased and blended at the beginning of the project and stored at UCR. To ensure that the quality of the biodiesel remains the same throughout the project, 200 ppmw of Tenox 21 (active ingredient is t-butyl hydroquinone) was added to the YGA manufactured by NFESC in Port Hueneme, California as recommended by NREL.

Since similar DoD vehicles may be operated under different conditions (i.e., different loads/routes) compared to the rest of a fleet, the testing program was developed to incorporate multiple test cycles/load points for each engine tested. Generally, multiple test cycles or load points were tested using the same dynamometer, test track or load bank.

3.6.2. Period of Operation

Testing was conducted over a period of approximately one year. The testing of each individual engine took between one day and five weeks depending on the test location and the number and complexity of the testing to be performed. Standard test cycles vary from approximately 1 hour to 3 hours; however, significant additional time was required for transportation of test engines or test equipment, equipment set-up, calibration, fueling, and preliminary analysis of test results.

3.6.3. Amount/Treatment Rate of Material to be Treated

The use of biodiesel has been reported to provide significant benefits in reducing criteria pollutants from the exhaust of diesel engines. Estimates of the expected reductions were previously provided in Table 2.1. The total quantity of pollutants produced by diesel exhaust is a function of their concentration and the volume of exhaust. Both the concentration of pollutants and the exhaust flow rate continuously change based on the engine's load, speed, and environmental factors. The exhaust stream also is directly proportional to engine horsepower. As an example, a 1994 model year Caterpillar CAT3516 engine rated at 2571 horse power has a full load exhaust flow rate of 14,417 cfm at a temperature of 940 degrees Fahrenheit. Such an engine, located at Naval Public Works Center, Norfolk, and currently fueled with low-sulfur diesel, produces NO_x emissions of 414.16 lbs/kgal of fuel, CO emissions of 117.7 lbs/kgal, HC emissions of 30 lbs/kgal, and PM emissions of 33.5 lbs/kgal. Using the average reductions given in Table 2.1, along with the criteria emissions identified above, one can gain an idea of the emissions reduction potential from implementing a B20 biodiesel fueling program.

3.6.4. Residual Handling

The technology to be demonstrated by this project does not generate any residual wastes that require disposal.

3.6.5. Operating Parameters for the Technology

The factor that makes biodiesel such a valuable clean fuel for the engine owner/operator is the fact that, generally speaking, biodiesel can be used in existing diesel engines without making any engine modifications. Because biodiesel is totally compatible with petroleum diesel in any percentage, it can be used as a direct replacement for petroleum diesel. Engine pressure and temperature operating conditions are generally very close between biodiesel and petroleum diesel. One concern with neat biodiesel (B100) is its cold-flow properties. These properties can however, be modified by additives, including the amount of petroleum diesel blended into the biodiesel.

3.6.6. Experimental Design

To verify the suitability of biodiesel in reducing air emissions from in-service DoD diesel engines, a comprehensive test program was developed. As shown in Table 3.2, the proposed test program included a wide variety of test engines, fuels, operating conditions and NO_x improvement additives. The project includes emissions testing for criteria pollutants as well as HAPs. Test methods approved by the EPA were used for applicable tests. A listing of the actual analytical testing methods is provided in Paragraph 3.7. To ensure data quality, testing using test cycles and static points were repeated, data points were recorded continuously during the tests and reported as the integrated result over the whole test period, and three testing organizations were employed. Testing results from each engine were compared with the previously completed testing and with similar work performed by other test programs.

As previously discussed, the diesel engines proposed for testing were selected based on a survey performed by NFESC for a completed ESTCP project. In this survey, DoD engines with high usage, as indicated by the number of similar engines/vehicles and by estimated hours of operation, were identified. For this project, emissions testing was performed on these high usage engines either in the field or at a laboratory. At a minimum, all engines were tested using a JP-8 or ULSD base diesel fuel and a B20 biodiesel fuel. Additional tests were performed using B20, B50, B70, and B100 biodiesel fuels manufactured from either soybean oil or yellow grease. In addition, one engine was selected for demonstrating the effectiveness of the EHN and DTBP cetane improvers in reducing NO_x emissions. These additives were chosen based on an investigation reported on in Reference 7.2.

Gaseous criteria pollutants including CO, HC, and NO_x were measured during all tests. The total weight of $_{PM2.5}$ emissions were measured for all engines tested by NREL and UCR. On a subset of engines, full chemical and physical characterization of the HAP and PM emissions were performed. Test results were reported in the form of emission factors and reported as grams per mile (g/mile), grams per gallon of fuel consumed or grams per brake horsepower hour (g/bhphr). The emissions tests were performed using a combination of various standardized stationary and transient driving test cycles, static and actual on-road testing. Emissions testing results reported in the scientific literature show that air emissions vary with a number of parameters, with the most important variables being the engine operating conditions. The testing conditions were chosen to come as close as possible to the expected certification or to representative in-use conditions as selected by other investigators for similar applications.

Since the purpose of our test program is to provide emissions factors for existing DoD engines installed in various types of DoD operated vehicles or equipment, all testing was performed with the engine installed in the applicable vehicle or equipment. Engine testing in an engine test cell was not part of the test program. Vehicle emission testing was performed either on-road or with the vehicle placed on a chassis dynamometer. Portable generators testing was performed using an electrical resistance load bank. The load bank was adjusted so that testing could be performed at various percentages of full engine load as specified in the EPA test method.

3.6.7. Demobilization

Following completion of the emissions testing, each of the test vehicles/stationary engines was returned to its owners. During the testing process, no engine modifications were made.

3.7. Selection of Analytical/Testing Methods 3.7.1 Selection of Analytical Methods

Emissions testing for this project was performed by three testing organizations, ATC, NREL and UCR. The type of data that was collected was previously identified in Table 3.2. The analytical testing instrumentation that was used is listed in Table 3.5. Although each testing organization employs similar analytical testing instrumentation, and utilizes similar analytical testing procedures specified in federal or recognized standard publications, they each have unique testing capabilities in terms of the types of tests that they can perform. These unique capabilities have been fully exploited by this project.

For the testing of regulated pollutants, emissions testing analytical test methods approved by the EPA, and found in the Code of Federal Regulations (CFR), were used. Specifically, testing was performed using the methods contained in 40CFR86 for control of emissions from new and in-use highway vehicles and engines. The detailed emissions test procedures for diesel engines are found in 40CFR86, Subpart N – *"Emission Regulations for New Otto-Cycle and Diesel Heavy-Duty Engines; Gaseous and Particulate Exhaust Test Procedures"* and more specifically in paragraph 86.1310 2007 *"Exhaust gas sampling and analytical system for gaseous emissions from heavy-duty diesel-fueled engines and particulate emissions from all engines."*

For the non-regulated emissions, the analysis methods are not found in the CFR. Instead these analyses were performed using industrial specifications and methods that are referenced in the scientific literature. The speciated C_{1-C12} volatile organic compounds (VOCs) were determined using methods developed in collaborative research between the automobile and petroleum industries under the Auto/Oil Air Quality Improvement Research Program (AO/AQIRP), as detailed in Reference 7.3. For the C_{1-C12} VOCs, sample collection was performed using Carbowwax/molecular sieve packed tubes and/or Tedlar bags followed by gas chromatography – FID analysis using a modified Auto/oil protocol. The tube sample collection procedure is discussed in greater detail in Reference 7.4. Aldehydes and ketone emission rates were collected using Dinitrophenylhydrazine (DNPH) cartridges and analyzed using a high-performance liquid chromatograph with ultraviolet detection, as per an AO/AQIRP method (Reference 7.3). Elemental Carbon/Organic Carbon samples were collected on quartz filters and analyzed using a Thermooptical carbon aerosol analyzer from Sunset Laboratories, a National Institute of Occupational Safety and Health (NIOSH) recognized method (References 7.5 and 7.6). Semi-volatile hydrocarbons were collected for analysis using a PUF/XAD cartridge immediately downstream of the quartz fiber media.

To detect gaseous air emissions in the laboratory, a non-dispersive infrared (NDIR) analyzer was used to measure CO and Carbon Dioxide (CO_2), a heated probe and flame ionization detector was used to measure HC's, and a chemiluminescence analyzer was used to measure NO_x. Portable versions of these instruments were available and were employed for field measurements. The mobile instrumentation used for this project used an NDIR for CO, CO_2, and HC and a solid-state zirconia sensor for NO_x.

Characterization of gaseous HAP compounds, including the Mobile Source Air Toxics identified in Table 3.6, were performed using Gas Chromatography (GC) where the samples were collected on DNPH cartridges. Acetaldehyde, Formaldehyde, Benzene, and 1,3-Butadiene are the 4 main gas-phase HAPs specified in the Clean Air Act for mobile sources. Acrolien is another gas-phase chemical targeted by EPA for its toxicity and ambient levels.

Naphthalene is the Polycyclic Aromatic Hydrocarbon (PAH) with the highest concentration in vehicle exhaust.

The measurement of PM emissions is more difficult and consisted of mass measurements as well as chemical characterization of the particles. Mass measurements were made by collecting particulates on a filter media and weighing the media before and after exposure to the exhaust. For these measurements, it is critical that the CFR methods be applied with respect to the use of an upstream classifier to remove the large particles and that the filter face temperature be maintained at 47°C ±5°C. Chemical characterization of the PM involved chemically testing the particles collected on quartz filter media for elemental and organic carbon as these measurements can be compared with similar data from ambient monitors to determine source signatures.

Table 3.5. Test Methods and Analysis of Exhaust Emissions

Instrument/Method	Measurement	Sample Duration	Lower Quantifiable Limit (Expressed in terms of fundamental measurement)
Pierburg NDIR	CO2, CO	1 s	50 - 500 ppm
California Analytical Instruments/Flame Ionization Detection	HC, Methane	1 s	10 - 30 ppm
California Analytical Instruments/Chemiluminescence	NO, NO2	1 s	10 ppm
Various/Filter*	PM2.5 Mass and Chemistry-	0.25 – 2 hrs	Various
Tedlar Bag/GC-FID	VOC's (C2 –C12)	0.25 – 2 hrs	10 ppb C
DNPH Cartridges/Shimadzu HPLC/UV	Aldehydes and Ketones	0.25 – 2 hrs	0.02 ug/mL

*Includes Teflon and quartz media for mass, metals, ions, elemental/organic carbon and PAHs by GC/MS on extracts from filters.

**Table 3.6. Partial List of EPA's Recognized
Mobile Source Air Toxics**

Acetaldehyde	Benzene	Formaldehyde
Acrolein	1,3-Butadiene	Naphthalene

The majority of the emissions testing program was performed by UCR utilizing their mobile heavy-duty testing laboratory (test trailer) (Reference 7.7). A schematic of the trailer is shown in Figure 3.1. This laboratory was designed for testing diesel powered generators and heavy-duty vehicles. The test trailer can be used for on-road tractor testing, testing of a generator connected to an electric load bank, or vehicle testing where the vehicle is placed on a separate chassis dynamometer. The UCR mobile laboratory dilutes the whole exhaust and utilizes the constant volume sampling concept of measuring the combined mass emissions of CO, CO_2, NO_x, Methane (CH_4), PM and Total Hydrocarbon (THC). Additionally, a proportional bag sampling for sample integration is used for HC, NO_x, CO, and CO_2 measurement. The mass of gaseous emissions is determined from the sample concentration and total flow over the test period. The mass of particulate emissions is determined from a proportional mass sample collected on a filter and the total flow over the test period.

For emissions measurements on light/medium duty vehicles, UCR has a Burke E. Porter 48-inch single-roll electric dynamometer. For emissions testing with this dynamometer, UCR utilizes standard bag measurements for CO, CO_2, NO_x, and THC. These measurements are conducted with a Pierburg AMA-4000 bench. THC is measured modally through a line heated to 190°C using a Pierburg AMA-2000 emission bench.

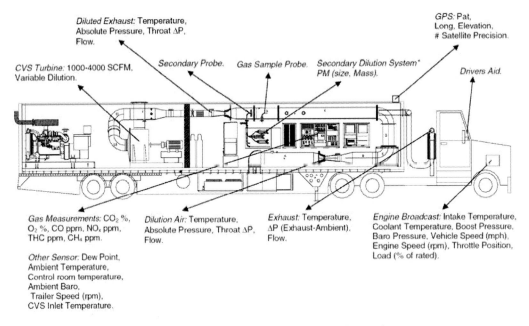

Figure 3.1. Schematic of the UCR Heavy-Duty Diesel Mobile Emissions Laboratory.

In-use testing was performed by ATC using the EPA's Real-time On-road Vehicle Emissions Reporter (ROVER). The ROVER system is presently used by ATC to perform tests for the EPA's program to monitor in-use heavy-duty diesel engines. The ROVER (including all of its components) is mounted on or in the vehicle. The data is recorded with the ROVER package computer on or in the vehicle. This data is monitored via antenna on another computer either onsite or in a chase vehicle. The main focus of the ROVER is measurement of NO_x emissions. It also records HC, CO, CO_2, exhaust temperature and pressures, torque, road speed, and the engine data stream. The engine data stream is contained in newer engines' computers systems. The ROVER system is comprised of two analyzers (both measure NO_x), a flow pipe (size determined by maximum flow of exhaust), a flow box, a pro-link tool (monitors the engine data stream), and computers containing the ROVER program. The flow pipe and box are attached directly to the engine exhaust. Sample ports in the flow pipe pull a sample and feed it to the analyzers, which are placed somewhere on or in the vehicle. The pro-link tool is set up with the analyzers and records the engine data stream to the ROVER computer. ROVER utilizes this tool with newer engines to monitor their data stream. This capability is only useful in newer vehicles containing the data stream computer system. The ROVER system records real-time data every second during the test run.

The NREL laboratory features a heavy-duty chassis dynamometer that simulates operation of a vehicle on the road. The dynamometer is connected with two 40-inch diameter rolls that are capable of testing all highway ready single or twin-axle vehicles. The distance between the rolls can be varied between 42 and 56 inches. The dynamometer will accommodate vehicles with a wheelbase between 89 and 293 inches.

In the NREL lab, simulations of vehicle loads including, rolling resistance, air resistance, desired road grade, and acceleration of vehicle inertia are performed with the dynamometer and controller software. Vehicles of weights between 8,000 to 80,000 lbs. can be simulated via electrical inertial simulation. For each vehicle test, standard or customized driving test cycles are used that match the duty-cycle of the test vehicle, ranging in speeds from idle up to 60 miles per hour. The dynamometer is equipped to run automated warm-up and coast-down routines to verify that dynamometer parasitic loads are stabilized and that road load simulations are accurate.

The NREL chassis dynamometer is supported by continuous exhaust emissions equipment similar to that previously described for the UCR Heavy-Duty Diesel Mobile Emissions Laboratory. An environmental chamber and microbalance specially designed to measure PM mass at EPA 2007 regulated levels is utilized. The lab does not, however, have the capability to chemically characterize HC and PM emissions to the extent of the UCR lab.

3.7.2. Selection of Testing Methods

3.7.2.1. Light/Medium-Duty Dynamometer Testing

The Humvee and Ford F350 were tested over the Federal Test Procedure (FTP) and US06 cycles for light-duty vehicles [Reference 7.8] using the UCR light/medium duty dynamometer. These vehicles were preconditioned prior to the first test on any new fuel by driving on the dynamometer over two back-to-back iterations of the LA4 driving schedule followed by an overnight soak at a temperature of approximately 72°F. Each vehicle was tested twice on each of the test fuels specified. A US06 cycle was run immediately after each FTP, with a preconditioning of 5 minutes at 50 mph to warm the engine up to operating temperature.

3.7.2.2. Heavy-Duty Dynamometer Testing

The F700 stakebed truck, F9000 truck, and a bus (Engine No. 8 of Table 3.2) were all tested over the AVL 8-Mode heavy-duty test [Reference 7.9]. This cycle is a steady-state test comprised of 8 modes under speeds and load ranging from idle to full load. The cycle was designed to closely correlate with the exhaust emission results over the US FTP heavy-duty engine transient cycle. The composite value is calculated by applying weighing factors to the results for the individual modes. The load points and weighting factors are provided in Figure 3.2. These vehicles were tested on a hydrostatic chassis dynamometer at a local Caterpillar dealer in Riverside, California. An engine map for the AVL 8-mode was conducted prior to initiating the testing on any of the fuel blends.

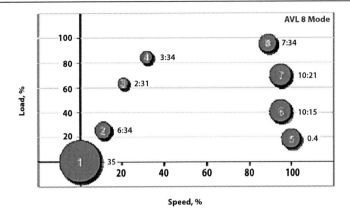

Figure 3.2. Load Points and Weighting for AVL 8-Mode Cycle.

Transient chassis dynamometer testing was conducted at the NREL ReFUEL laboratory on a Cheyenne Mountain Air Force Base operated bus. The transient test was a special cycle designed to specifically simulate operation of the buses within Cheyenne Mountain. The Cheyenne Mt. Cycle is shown in Figure 3.3. This cycle was developed based on activity data monitored from actual buses operating with the Cheyenne Mt. Facility. It is composed of 6 primary events where the vehicle is accelerated to a speed of between 20 to 32 mph. A total of 2.5 miles are driven over a 1,200 second duration.

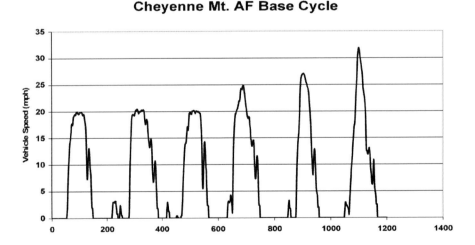

Figure 3.3. Speed vs. Time Traces for the Cheyenne Mountain Cycle.

3.7.2.3. Portable Generator Testing

Testing for the portable generators was performed over a five-mode test cycle at steady-state conditions, as described in Table 3.7. This is a standard certification cycle for testing non-road diesel engines that is described in EPA's 40CFR Part 89 [Reference 7.10] and by the International Standards Organization (ISO) [Reference. 7.11]. While both EPA and ISO testing procedures are the same, the analysis of the results differ in that the EPA method, which was used in this study, only corrects the NO_x whereas the ISO applies a correction factor for moisture to both the PM and NO_x. The standard test protocol consists of a series of preconditioning cycles to warm and stabilize the engine followed by a sequence of stabilization and testing at five modes, each with a defined speed and load. The engine is run at rated speed for a minimum period while measuring the regulated emissions. The engine is preconditioned at idle and then full power for at least 30 minutes before measurements are made. Testing begins at the 100% mode and moves from there to the lower power modes with measurements collected for at least 10 minutes at each mode. For the duplicate run, the whole procedure is started over from the beginning. Emissions from the portable generators were measured using the UCR Heavy-Duty Laboratory.

Table 3.7. Five-Mode Test Cycle for Constant Speed Engines

Mode number	Engine Speed[1]	Observed Torque[2]	Minimum time in mode, min.	Weighting factors	Mode number
1	Rated	100	5.0	0.05	1
2	Rated	75	5.0	0.25	2
3	Rated	50	5.0	0.30	3
4	Rated	25	5.0	0.30	4
5	Rated	10	5.0	0.10	5

Notes:
1. Engine speed: ± 2 percent of point
2. Torque: Throttle fully open for 100% point. Other points: ±2% of engine maximum.

3.7.2.4. In-Use Testing

In-use tests were conducted on three pieces of equipment at ATC. This included a forklift, airport tow vehicle, and a Humvee. For the forklift, the test runs simulated forklift usage include idling, hydraulic usage while idling, and driving with and without hydraulic usage over a 3.6 mile test run. In the past 2 years of EPA non-road testing, it was determined that loading the engine to simulate use of vehicles' hydraulics can be maximized by "dead heading" the hydraulics. In this case, that involves simply running the forklift's forks in a certain direction, i.e., pulling them back until they won't move anymore and holding that position. During this test, this was used to simulate the load of using the forklift's hydraulics.

The test run for the forklift with both fuels was as follows:

1) Idle time: approximately 5 minutes w/out load on engine
2) Dead head hydraulics at idle: 5 minutes of 30 seconds dead heading hydraulics while idling and 30 seconds idle w/out (normal low idle)
3) Dead head hydraulics at higher idle: 5 minutes of 30 seconds dead heading hydraulics while idling at higher RPM and 30 seconds idle w/out
4) Run included repeatable 3.6 miles with seven different periods of 15 second dead heading hydraulics at same locations on each run
5) Idle time: after run another 5 minutes of 30 seconds dead heading and 30 seconds of idle (normal low idle)

For the aircraft tow vehicle, test runs included driving the aircraft tow on the same route around the APG airfield, achieving maximum speed at the same point on each run. Each run was approximately 1.4 – 1.6 miles. The Humvee test runs were designed to simulate normal driving in fleet usage. The Humvee was operated leaving APG and driving west on Maryland RTE 40 to White Marsh, MD and back to APG. This route is stop and go due to traffic lights on Maryland RTE 40. Continuous usage with stop and go is the best representation of APG fleet usage.

3.8. Selection of Analytical/Testing Laboratory

As previously described, the emissions testing will be performed on-site and in laboratories operated by NREL and UCR. All required analytical testing of the fuels will be performed at UCR or at commercial testing laboratories under contract to UCR.

4. PERFORMANCE ASSESSMENT

4.1. Performance Criteria

Table 4.1. ESTCP Performance Criteria

Performance Criteria	Description	Primary or Secondary
Criteria Air Pollutant Emissions	Reduce CO, HC & PM Air Pollutant Emissions, Minimizes NOx Emission Increases	Primary
HAP Emissions	Reduce HAP Pollutant Emissions	Primary
NOx Reduction Additive	Reduce NO_x emissions	Secondary
Fuel Economy	Maintain Fuel Economy Consistent with Energy Content of Fuel	Secondary
Drivability	Maintain Engine Performance	Secondary
Reliability	No Maintenance Increase	Secondary

4.2. Performance Confirmation Methods

Since the purpose of this demonstration is to obtain air emissions data for DoD diesel engines of interest that is not currently available in the literature, the overall success of this project will be measured in terms of the quality of data acquired and its acceptance by the scientific community. As an additional measure, this project must provide sufficient data to convince DoD diesel fleet operators and fuel suppliers to implement B20 biodiesel programs within their activities. In order for the project's test results to be accepted, standard recognized test methods must be employed and the results reported in units consistent with other investigations. In addition, the results will be compared with other previous investigations and with the emissions models provided by the EPA (see Reference 7.1).

For this project, standard EPA approved test methods have been used for both the laboratory and field measurements. To ensure the engines are consistently loaded, a chassis dynamometer or an electric load bank was used during the majority of the testing. All testing, except for those performed on-road, were performed with the engines operating on standard test cycles. All test cycles were repeated with the reported results being the average of the tests. Gaseous air emissions data was continuously measured over the test cycle, with the results reported as an integrated value. Particulate emissions were collected on a filter paper throughout a cycle and weighed after the testing is complete. The testing equipment, as previously described in paragraph 3.7 was used. All testing organizations participating in this project have extensive experience performing emissions testing. Their results from previous test efforts have been widely published in the literature.

Expected and actual engine performance from the demonstration and the applicable performance confirmation methods are shown in Table 4.2. Since emission testing provides quantitative results, this project will not have any primary qualitative performance objectives.

Table 4.2. Expected Performance and Performance Confirmation Methods

Performance Criteria	Expected Performance (pre demo)	Performance Confirmation Method	Actual (post demo) (future)
Primary Criteria (Performance Objects) (Quantitative)			
Reduce CO Emissions	Reduce emissions by 9% (min.) with B20	40 CFR 86	No change
Reduce HC Emissions	Reduce emissions by 16% (min.) with B20	40 CFR 86	No change
Reduce PM Emissions	Reduce emissions by 8% (min.) with B20	40 CFR 86	No change
Reduce HAP Emissions	Reduce emissions by 16% (min.) with B20	Various EPA methods	No change
Secondary Performance Criteria (Quantitative)			
Minimize increase in NO_x emissions	Emissions increase <3% for B20 (Without additive)	40 CFR 86	No change
Minimize increase in NO_x emissions	Reduce emissions by 2% (min.) for B20 (With additive)	40 CFR 86	No change
Fuel Economy	Similar to petroleum diesel	40 CFR 86	No change
Secondary Performance Criteria (Qualitative)			
Driveability	No change	Driver response	No change
Reliability	No change	Driver response	No change

4.3. Data Analysis, Interpretation and Evaluation

In this section, the emissions results for the multiple DoD operated mobile and portable diesel engines fueled with various types of biodiesel will be reported and discussed. The results are broken down for the purpose of this section into major groups corresponding to the type of diesel engine or its application. In order to present the CO, HC, NOx and PM results on the same graph, it was required to multiply the CO and HC results and divide the PM results. On each graph, the multiplication and division factors are identified. As an example, CO*5 indicates that the CO emissions factor should be multiplied by 5. Regulated and unregulated emissions are also discussed separately.

In addition to the collection of emissions data, Table 4.2 also identifies as Secondary Performance Criteria, the collection of fuel economy, drivability and reliability information. Based on energy content data reported in Reference 7.1, the project team did not expected that any fuel economy differences would be observable between the USLD, YGA and JP-8 fuels. This expectation proved to be correct. The fuel analysis reported in Tables 3.3 and 3.4 showed that the ULSD had a 1.6 percent higher energy content than the YGA fuel and the JP-8 fuel had a 0.9 percent higher energy content. These differences were less than the expected 3-5 percent difference.

For the drivability and reliability performance criteria, information was collected from fleet and vehicle maintenance management personnel at Camp Pendleton as well as from the emissions testing drivers. Based on interviews of these personnel, the project team concluded that vehicle drivers and maintenance mechanics experience no difference when operating or repairing B20 fueled vehicles. This result matched our expectation.

4.3.1. Regulated Emissions

4.3.1.1. Light/Medium-Duty Vehicles

The regulated emissions result for the 2004 Humvee and the 1999 Ford F-350 pick-up truck are presented in Figures 4.1 and 4.2 for the FTP and Figures 4.3 and 4.4 for the US06 cycle. These data represent the average of all tests conducted for each vehicle/fuel combination, with the error bars representing the standard deviation of the emissions tests.

PM emissions showed some trends with the different fuels for the Humvee over the FTP. The biodiesel blends generally showed reductions in PM. For the blends from B50 to B100, FTP PM reductions ranged from 23-42%. The B20 blends did not show as significant reductions, with the B20-YGA showing no reductions relative to the ULSD and the B20 Soy showing PM reductions of approximately 10%. PM emissions increased for the JP-8 fuel for the Humvee by approximately 10%. For the F350 over the FTP, no statistically significant differences in PM were found between the ULSD and the B20-YGA fuels.

For the more aggressive US06 cycle for the Humvee, PM emissions for nearly all fuels showed reductions relative for the ULSD. The ULSD and JP-8 PM results showed significant variability as shown by the large error bars, however, in a number of cases, the differences were within the experimental error. It is possible that more aggressive preconditioning than that used in this study might be needed to obtain more stable PM readings over the US06. The lowest overall PM emissions over the US06 were found for the higher biodiesel blends (i.e., 50-100%), consistent with the FTP results. For the F350, the B20 YGA PM results were approximately 10% lower than for ULSD.

Emissions of NO_x did not change significantly over the range of fuels tested on the Humvee and F-350. The general lack of trends in NO_x emissions was consistent between the FTP and US06 test cycles. For the biodiesel blends, NO_x emissions were comparable with those of the ULSD within experimental variability, showing no NO_x disadvantage. The additives also did not show a strong affect on NO_x emissions compared to the baseline neat biodiesel tests. On the Humvee, slight increases in NO_x were observed for the JP-8 over both the FTP and US06 cycles. For the F350, NO_x emissions for the ULSD and B20-YGA were all comparable within the experimental variability.

THC emissions showed different trends between the two vehicles. For the Humvee, the ULSD fuel provided the lowest THC emissions of all of the fuels tested, with most fuels having 75-130% higher THC emissions over the FTP. The JP-8 showed the largest increase in THC over the FTP, with increases relative to ULSD of ~250% for the FTP. Over the US06 for the Humvee, JP-8 showed an ~80% increase in THC relative to the ULSD. There was a tendency toward higher THC emissions for the higher blend levels of biodiesel also on the US06, but these results were not statistically significant.

CO emissions also showed different trends with fuels for the two vehicles. For the Humvee, CO emissions were the lowest for the ULSD over both the FTP and US06. For the Humvee FTP tests, most fuels showed approximately a 15-30% increase in comparison with the ULSD. The increase in CO emissions for most fuels were slightly greater over the US06 relatively to the ULSD, with increases of between 20-60%. The highest CO emissions for both the FTP and US06 cycles for the Humvee were with the JP-8 fuel (110-130% higher than the ULSD). For the F-350, CO emissions showed a slight decrease over the FTP and no difference over the US06 for the B20-YGA.

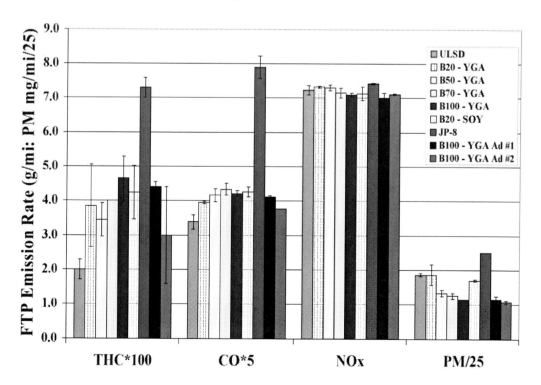

Figure 4.1. UCR-FTP Emissions Results 2004 Humvee.

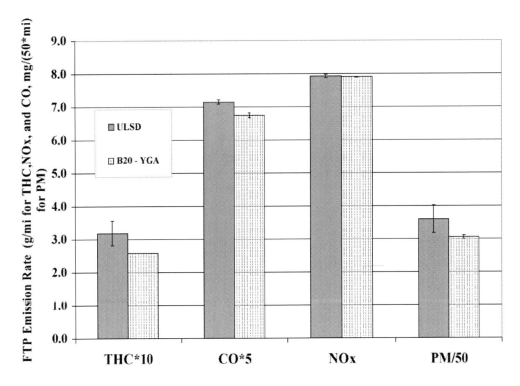

Figure 4.2. UCR-FTP Emissions Results 1999 Ford F350 Pick-Up Truck.

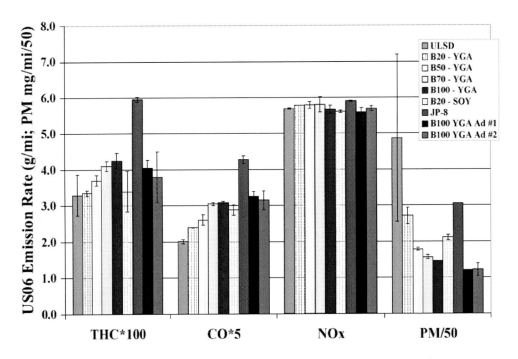

Figure 4.3. UCR - US06 Emissions Results 2004 Humvee.

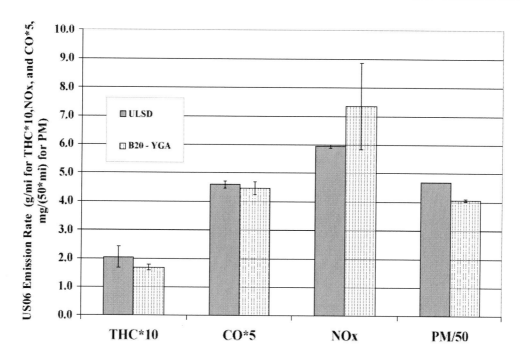

Figure 4.4. UCR-US06 Emissions Results 1999 Ford F350 Pick-Up Truck.

4.3.1.2. Heavy-Duty Vehicles

Fuels effects for the heavy-duty vehicles differed depending on the specific test vehicle. The results for the Ford F9000, Ford F700, and Pendleton Thomas bus over the AVL 8 mode cycle are shown in Figures 4.5-4.7. The results for the Cheyenne Mountain Thomas bus over the Cheyenne Mountain Cycle are shown in Figures 4.8 and 4.9, respectively, for measurements by UCR and NREL.

4.3.1.2.1. AVL 8-Mode Results

For the Ford F9000, reductions in PM relative to the ULSD were found for the B20-YGA, and B20-YGB fuels of between 20-40%. Reductions of 5-15% were also found for CO for B20-YGA and B20-YGB. The F9000 showed a trend of 30-40% lower THC and CO emissions with the B100 fuel, about a 70% reduction in PM, and a 41% increase NO_x emissions over the AVL 8- mode cycle. The Ford F700 AVL-8 mode results did not show any significant differences between the ULSD and the B20-YGA and B20-YGB fuels. The Pendleton bus over the AVL 8- mode did not show fuel differences that were statistically significant.

In summarizing the AVL 8-mode results, in a number of cases, the differences in the fuels were small for most of the emissions components. The biodiesel blends showed some reductions in PM for the F9000, but not for the other vehicles. Some slight reductions in CO were also found for the F9000. Although the B100 was tested on only one vehicle, PM, THC, and CO emissions decreased while NO_x increased, consistent with general trends for biodiesel.

4.3.1.2.2. Cheyenne Mountain Results

The trends in the Cheyenne Mountain Test results showed some differences between the measurements made by UCR and NREL. The UCR and NREL measurements were performed on the same dynamometer, with the same cycle and driver, but they were sampled at different times during the day. The results for UCR showed reductions in PM and CO for the B20-Soy compared with the ULSD. The results for the NREL measurements, on the other hand, showed no differences between the PM and CO emissions on B20-Soy and ULSD. The differences in the PM results could be due to differences in the temperatures at which the PM is measured. The UCR – Mobile Emissions Laboratory (MEL) measures PM at 47°C ±5°C, whereas the NREL PM measurements are made at room temperature. At the higher measurement temperature, the UCR PM samples would be expected to have less volatile compounds. The NREL measurements showed reductions in THC with the B20-Soy. THC emissions for the B20-soy were also lower than the ULSD for the UCR – MEL, however, this result was not statistically significant. Both UCR and NREL showed no statistically significant differences in NO_x emissions between the B20-soy and ULSD.

1.3.1.2.3. Portable Generator Results

The results for the 250 kW generator and a 60 kW tactical generator are shown in Figures 4.10 and 4.11, respectively. Of the fuels tested on both generators, the JP-8 showed increases in THC and CO for both generators. The JP-8 also showed about a 50% reduction in PM for the 60 kW generator. The B100 blend showed a reduction in THC for the 250 kW generator, but no other significant emissions effects. The B20 YGA showed little significant changes in any of the emissions components for either of the generators tested, with only a slight reduction in THC found for the 60 kW generator.

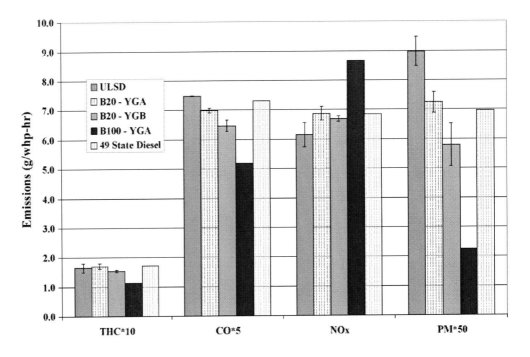

Figure 4.5. UCR-Chassis AVL 8 Mode Results Ford F9000.

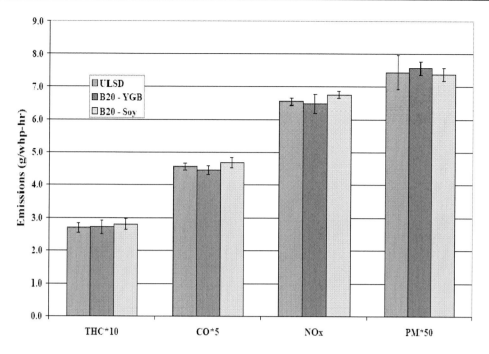

Figure 4.6. UCR-Chassis AVL 8 Mode Results Ford F700 Stakebed Truck.

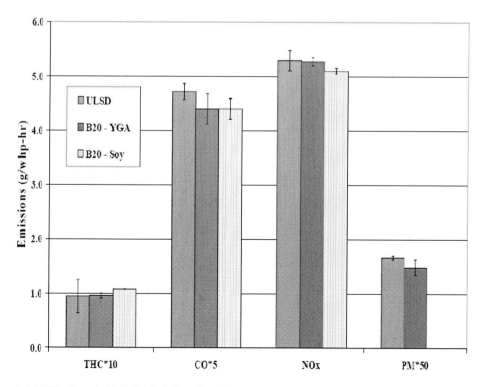

Figure 4.7. UCR-Chassis AVL 8 Mode Bus Results.

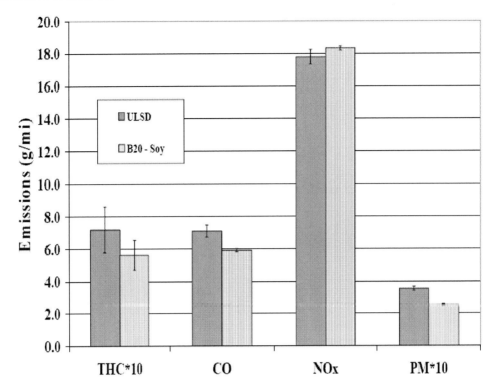

Figure 4.8. UCR-Cheyenne Mountain Cycle Results – Cheyenne Mountain Bus.

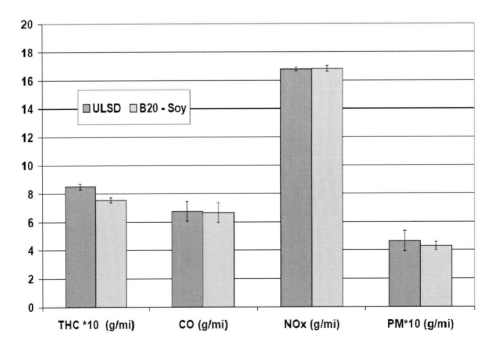

Figure 4.9. NREL-Cheyenne Mountain Cycle Results – Cheyenne Mountain Bus.

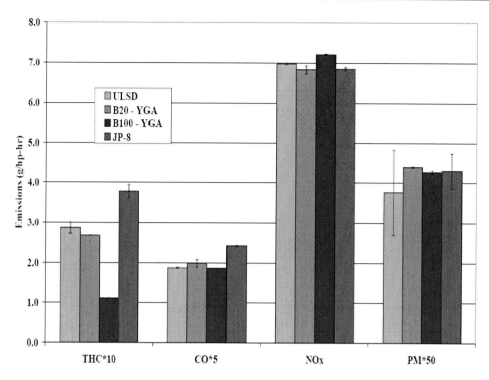

Figure 4.10. UCR – 5-Mode Test Results – 250 kW Generator.

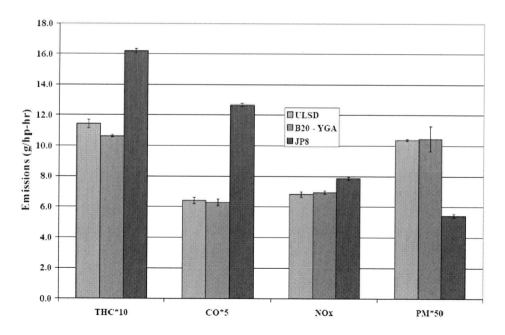

Figure 4.11. UCR – 5-Mode Test Results – 60 kW Generator.

4.3.1.2.4. In-Use Results

In-use testing results for a Hyster forklift, a Harlan aircraft tow, and a Humvee are shown in Figures 4.12 through 4.14, respectively. For the forklift, approximately a 20% increase in CO emissions was found for the B20 soy-based biodiesel fuel. As only a single test iteration is available, it is uncertain if this difference was significant or not. The differences for THC and NO_x for the forklift were small and likely within the experimental variability. For the aircraft tow vehicle, the differences between the ULSD and the B20 – YGA were all within the experimental variability. For the Humvee, only a single test iteration was available which showed little difference between the ULSD and the B20-Soy for CO and NO_x. THC was not available for the Humvee because a heated HC line was not used for these tests.

4.3.1.2.5. Discussion – Regulated Emissions

The primary fuels of interest for this study were the B20 biodiesel blends, since this is blend of biodiesel used in military vehicles. The project results for the regulated emissions were that at the B20 level, there were no consistent trends over all applications tested. Within the context of the test matrix, no differences were found between the different YGA, YGB, and Soy-based biodiesel feedstocks. The results of more extensive statistical analyses also indicated no statistically significant differences in CO, HC, NOx and PM emissions between the B20-YGA and the ULSD. The tested NOx reduction additives also proved to be ineffective. Thus the air pollution performance objectives outlined in the project's demonstration plan were not met. Although these results were not expected, they are not necessarily a disappointment since the baseline USLD fuel proved to be greatly superior to existing on-road Diesel No. 2.

The project results showed that over the range of vehicle/equipment types, emission factors could significantly vary depending on application or type of usage. A comparison of the emissions differences is provided in Table 4.3 for all vehicle/fuel combinations. Statistical comparisons between the different fuels and the ULSD using a standard t-test are also provided in Table 4.3. For this analysis, we considered $p \le 0.05$ as statistically significant and $p \le 0.10$ as being marginally statistically significant.

Although there were no overall trends, there were trends for individual engines. For the Ford F9000 tractor, there was a trend of lower PM emissions for the B20- YGA and B20-YGB fuels. There was some trend of higher HC emissions with the biodiesel blends for the Humvee, considering also the higher blend levels. The B20-YGA and B20 YGB also showed a trend of higher CO emissions on the Humvee.

To provide a better understanding of the effects of B20 over the entire fleet, some additional statistical analyses were performed. Since the vehicles/equipment represent a variety of applications and test protocols, the results were normalized into units of grams of emissions per either gallons of fuel used, kg of fuel used, or BTU's of fuel used. This provides a mechanism for which the results of the fleet as a whole could be compared on a consistent basis. These analyses were performed comparing ULSD to B20-YGA, since this was the blend utilized with nearly all of the test vehicles. For this analysis, a two-tailed, paired t-test was performed using only the average values for a particular vehicle/fuel combination. As such, this analysis does not account for the variability of testing within a specific vehicle/fuel combination. The results for HC, CO, NO_x and PM all showed no statistically significant differences between the ULSD and the B20-YGA for either the calculations based on gallons of fuel used, emissions per kg of fuel used, or BTU of fuel used.

Statistical comparisons for fuel consumption were also made with the results showing no statistically significant difference in fuel consumption between the ULSD and the B20-YGA. A summary of these statistical analysis results is provided in Table 4.4.

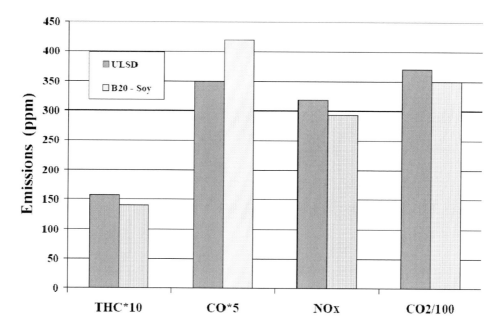

Figure 4.12. ATC – Hyster Forklift Emissions Results.

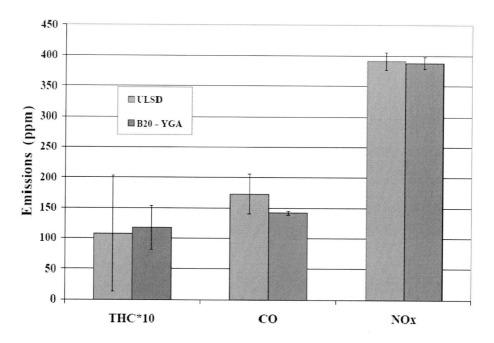

Figure 4.13. ATC – Aircraft Tow Emissions Results.

Table 4.3. Summary of Emissions Changes and Statistics Relative to ULSD for Individual Vehicles

Vehicle	Cycle		YGA - B20 HC	CO	NOx	PM	Soy - B20 HC	CO	NOx	PM	YGB - B20 HC	CO	NOx	PM	JP-8 HC	CO	NOx	PM	YGA - B100 HC	CO	NOx	PM
F350	FTP	% change	-19%	-6%	0%	-15%																
		p-value	0.15	0.03	0.65	0.21																
	US06	% change	-18%	-3%	24%	-13%																
		p-value	0.31	0.53	0.32	NA																
Model A2 Humvee	FTP	% change	93%	17%	1%	0%	113%	26%	-1%	-9%					265%	133%	3%	35%	133%	24%	-2%	-38%
		p-value	0.17	0.06	0.48	0.99	0.06	0.04	0.61	0.05					0.00	0.00	0.20	NA	0.03	0.04	0.31	**
	US06	% change	2%	19%	2%	-44%	3%	44%	-1%	-67%					80%	113%	4%	-37%	29%	54%	0%	-70%
		p-value	0.91	0.01	0.01	0.32	0.88	0.01	0.09	0.23					0.02	0.00	0.01	NA	0.16	0.00	0.80	NA
F700	AVL 8-mode	% change					4.8%	2.7%	3.2%	5.6%	0.4%	-2.1%	-0.9%	-5.4%								
		p-value					0.14	0.33	0.14	0.90	0.96	0.63	0.40	0.81								
F9000	AVL 8-mode	% change	3.2%	-5.7%	11.7%	-9.4%					-0.5%	-3.5%	8.6%	-5.5%					-1.6%	-0.6%	40.8%	-4.9%
		p-value	0.72	0.01	0.16	0.05					0.21	0.02	0.22	0.04					**	**	**	**
Camp Pendleton Bus	AVL 8-mode	% change	1.3%	-5.8%	-0.4%	-0.8%	13.3%	-0.7%	-0.7%													
		p-value	0.96	0.29	0.91	0.28	0.62	0.20	0.29													
250 kW Generator	5-mode	% change	-0.6%	5.8%	2.3%	17.1%									31.8%	28.9%	1.5%	14.3%	-1.4%	-0.7%	8.3%	13.3%
		p-value	0.20	0.25	0.21	0.48									0.03	0.00	0.11	0.58	0.00	0.21	0.00	0.58

Vehicle	Cycle		YGA - B20				Soy - B20				YGB - B20				JP-8				YGA - B100			
			HC	CO	NOx	PM	HC	CO	NOx	PM	HC	CO	NOx	PM	HC	CO	NOx	PM	HC	CO	NOx	PM
60 kW generator	5-mode	% change	6.3%	13.0%	8.2%	10.9%									42.0%	97.5%	15.6%	-18.1%				
		p-value	0.69	0.41	0.40	0.57									0.00	0.00	0.02	0.00				
Cheyenne Mountain Bus – LICR	Custom	% change					-22.0%	-17%	3.0%	-29%												
		p-value					0.18	0.01	0.12	0.00												
Cheyenne Mountain Bus – NREL	Custom	% change					-1.2%	-3%	0.2%	-3.4%												
		p-value					0.00	0.87	0.81	0.39												
Aircraft tow	In-use	% change	9%	-18%	-1%																	
		p-value	0.91	0.31	0.83																	
Forklift	In-use	% change					-10%	20%	-8%													
		p-value					**	**	**													
Model A1 Humvee	In-use	% change	NA	5%	6%																	
		p-value	**	**	**																	

Table 4.3 (Continued)

Summary of Emissions Changes and Statistics Relative to ULSD for Individual Vehicles

Vehicle	Cycle		YGA - B50				YGA - B70				YGA - B100 + Additive 1				YGA - B100 + Additive 2			
			HC	CO	NOx	PM	HC	CO	NOx	PM	HC	CO	NOx	PM	HC	CO	NOx	PM
Humvee	FTP	% change	73%	23%	1%	-28%	100%	28%	-1%	-33%	120%	21%	-3%	-38%	50%	11%	-2%	-42%
		p-value	0.07	0.06	0.62	0.02	0.01	0.04	0.63	0.01	0.01	0.04	0.24	0.01	0.43	0.13	0.34	0.00
	US06	% change	12%	30%	2%	-63%	24%	52%	2%	-68%	23%	62%	-2%	-75%	15%	57%	0%	-75%
		p-value	0.43	0.03	0.30	0.20	0.19	0.00	0.55	0.18	0.22	0.01	0.37	NA	0.52	0.03	0.99	0.16

** Insufficient data for t-test calculation

95% confidence level - statistically significant - $p \leq 0.05$

90% confidence interval - marginally statistically significant - $\leq 0.05\ p \leq 0.10$

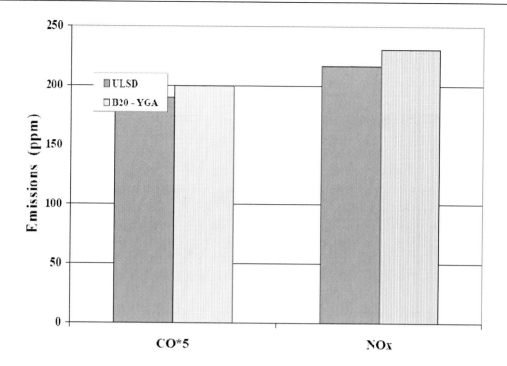

Figure 4.14. ATC – Humvee Emissions Results.

Table 4.4. Fleetwide Statistical Analysis Results for B20-YGA vs. ULSD

		HC	CO	NO$_x$	PM	Fuel Use
Emissions per kg of fuel used	%	-8.6%	-5.1%	0.0%	-9.2%	
	p-value	0.14	0.28	0.97	0.27	
Emissions per gal. of fuel used	%	-7.6%	-4.1%	+1.1%	-8.2%	
	p-value	0.16	0.36	0.19	0.32	
Emissions per BTU of fuel used	%	-7.3%	-3.8%	+1.4%	-7.9%	
	p-value	0.16	0.39	0.11	0.33	
Gallons per work or activity unit						+1.7%
						0.18

The higher biodiesel blends (B50 to B100) were only tested on the Humvee, and with the B100 on the 250 kW generator and a single test on the Ford F9000. On the Humvee, the higher biodiesel blends did show a trend of higher CO emissions with the higher biodiesel blends, consistent with the B20 blends. There was also a general trend of higher HC emissions, at least on the FTP, for this vehicle. Finally, there were some trends of lower PM emissions on the Humvee for the higher biodiesel blends and on the F9000 for the B100. This is consistent with the larger body of literature, although consistent PM reductions are not found at the B20 blend levels. NO$_x$ emissions for the single test on the F9000 with B100-YGA were also higher than that found for ULSD.

JP-8 was also tested over a range of test applications. The JP-8 showed relatively consistent higher HC emissions over the Humvee and the two generators ranging from 30 to 265%. Similarly, CO emissions increased with JP-8 on the Humvee and the two generators in the range of 30 to 130%. Some improvement in PM was found with the JP-8 on 60 kW generators.

The results of this study in general show much smaller changes in emissions with B20 than previous studies. A number of other studies have found larger reductions in HC, CO, and PM emissions for biodiesel fuels [References 7.1 & 7.12 - 7.20]. There are some differences, however, between the present and previous studies. In many of the previous studies, however, comparisons were made with Federal No. 2 diesel fuels with higher aromatic contents and lower cetane numbers than the CARB fuel used in the present work [References 7.12-7.18]. Other previous studies have, however, demonstrated the emissions reduction potential of biodiesel blends in comparison with CARB fuels [References 7.19, 7.20]. It is worth noting that while ULSD will soon be implemented throughout the country, the nature of the diesel fuel could still differ significant between different regions of the country. The CARB ULSD probably represents the most stringent fuel requirements that would be met by commercial fuels. In other parts of the country where fuel specification on aromatics and other fuel properties are not as strict, some additional benefits may be found relative to those in this study.

There may also be differences in the operational load in comparison with engine dynamometer tests that affect the magnitude of the changes in emissions. Previous studies have shown that the benefits of biodiesel fuels decrease in magnitude at lower loads [References. 7.18, 7.21 & 7.22]. UCR has also observed similar results for previous tests conducted over the light-duty FTP in their laboratory for biodiesel fuels on medium-duty diesel trucks [References 7.23 & 7.24].

4.3.2. Unregulated Emissions

4.3.2.1. Elemental and Organic Carbon

For mobile sources, the most significant component of the PM is the elemental and organic carbon. Figures 4.15 and 4.16 provide the elemental and organic carbon data for the Camp Pendleton bus and the Ford F9000 tractor. The data for the larger engines are provided on a CO2 basis so that comparison between vehicles can be more readily made. Also, the data are multiplied by the weighting factor to show the relative contribution on a basis of cycle weighting.

For mobile sources, the most significant component of the PM is the elemental and organic carbon. Figures 4.15 and 4.16 provide the elemental and organic carbon data for the Camp Pendleton bus and the Ford F9000 tractor. The data for the larger engines are provided on a CO2 basis so that comparison between vehicles can be more readily made. Also, the data are multiplied by the weighting factor to show the relative contribution on a basis of cycle weighting.

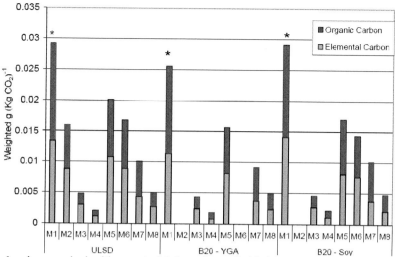

Note: Mode 1 values marked with an asterisk have been divided by 5 in order for other modes to be visible on the graph.

Figure 4.15. Bus Organic and Elemental Carbon Emission Rates Weighted by Mode.

Comparing across all fuels and tests, the ratio of elemental to organic carbon varies considerably depending on the vehicle, test mode, and fuel. Mode 1 of the AVL 8-mode cycle makes the largest contribution to the weighted average and is approximately equally weighted between elemental and organic carbon for most tests. For the F9000 operating on ULSD and YGA B20 fuel, organic carbon represented the largest fraction of the PM. On the other hand, the YGA B100 shows a larger fraction of elemental carbon on nearly all modes. Figure 4.17 and Figure 4.18 provide the relative contribution of Elemental Carbon (EC) and Organic Carbon (OC) to total Carbon (TC) for the bus and F9000, respectively.

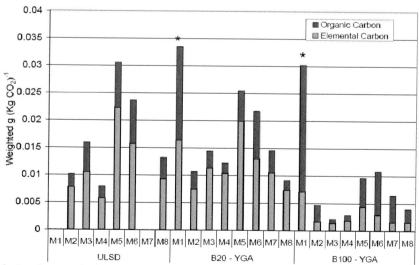

Note: Mode 1 values marked with an asterisk have been divided by 5 in order for other modes to be visible on the graph.

Figure 4.16. F9000 Organic and Elemental Carbon Emission Rates Weighted by Mode.

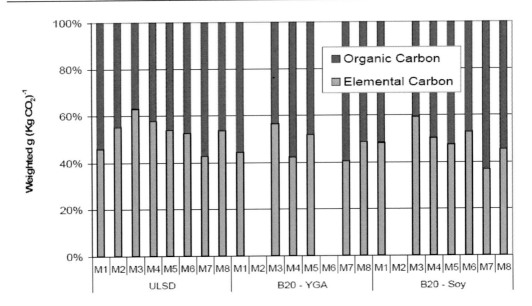

Figure 4.17. Bus Relative Organic and Elemental Carbon Emission Rates Weighted by Mode.

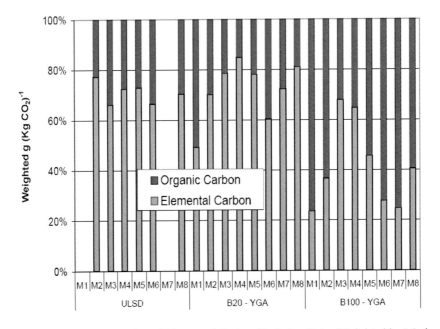

Figure 4.18. F9000 Relative Organic and Elemental Carbon Emission Rates Weighted by Mode..

4.3.2.2. Carbonyls

Figures 4.19, 4.20, and 4.21 present the weighted modal emission factors for carbonyl compounds formaldehyde, acetaldehyde, and acrolein for the Camp Pendleton bus, Ford F9000 tractor and the 250 kW generator, respectively. Figure 4.22 presents the FTP weighted carbonyl emissions for the Humvee. From these charts, it can be seen that mode 1 for the AVL 8-mode cycle is the most significant contributor to carbonyl emissions, similar to that seen for the organic carbon and elemental carbon. Mode 8 provides the second most significant contribution to weighted carbonyl emissions. For the portable generator 5-mode cycle, a more even distribution of the carbonyl compound emissions by mode is seen.

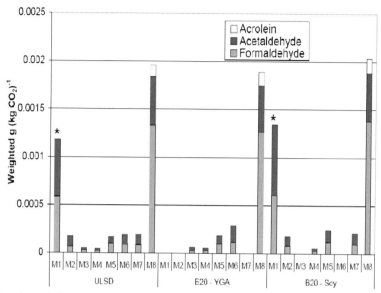

Note: Mode 1 values delineated with an asterisk have been divided by 5 in order for other modal emissions to be visible on the graph.

Figure 4.19. Bus Carbonyl Emissions Weighted by Modes.

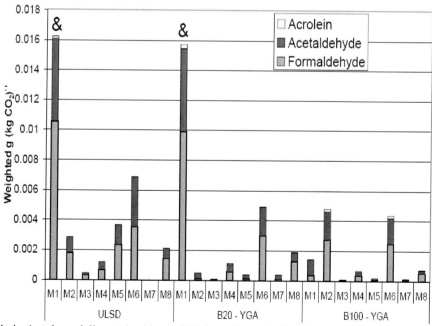

Note: Mode 1 values delineated with an "&" have been divided by 10 to allow for other modal emissions to be visible.

Figure 4.20. F9000 Carbonyl Emissions Weighted by Modes.

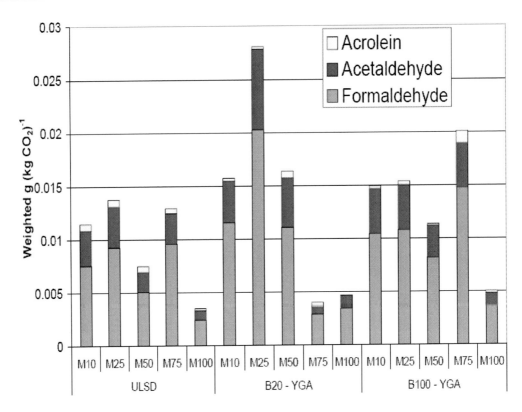

Figure 4.21. 250 kW Generator Carbonyl Emissions Weighted by Modes.

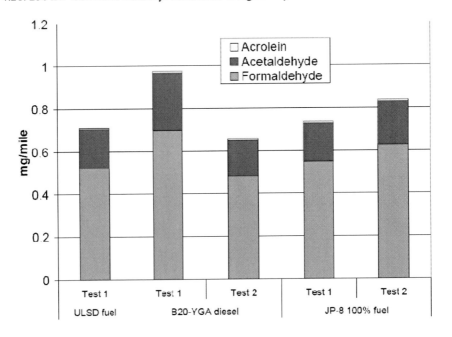

Figure 4.22. Humvee Carbonyl Emissions Over Weighted FTP Cycle in Grams Per Mile.

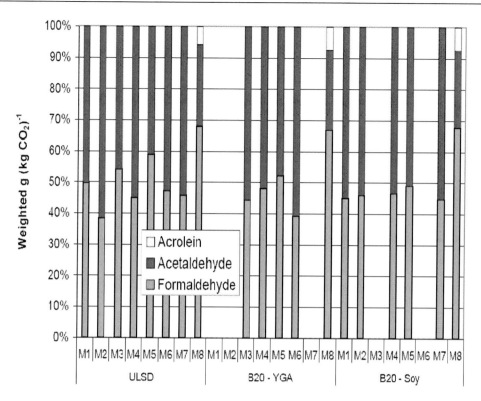

Figure 4.23. Bus Relative Carbonyl Emissions.

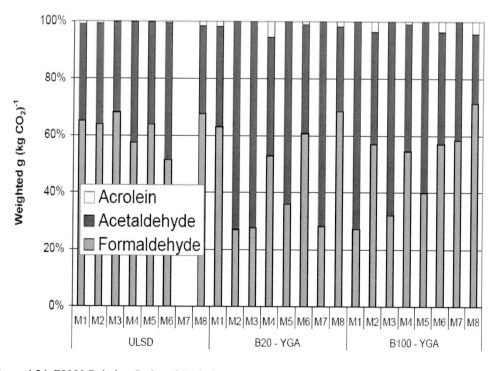

Figure 4.24. F9000 Relative Carbonyl Emissions.

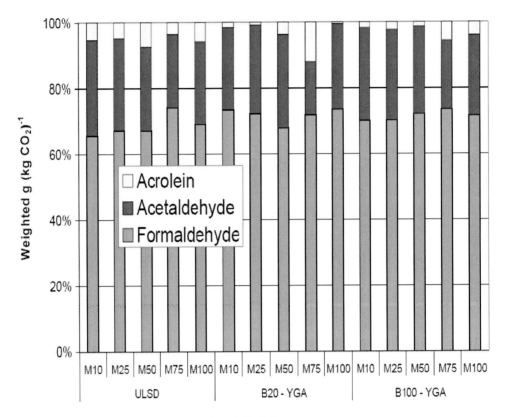

Figure 4.25. 250 kW Generator Relative Carbonyl Emissions.

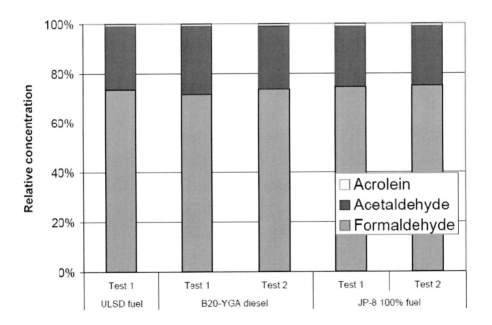

Figure 4.26. Humvee Relative Carbonyl Emissions.

Figures 4.23, 4.24, 4.25, and 4.26 provide the relative contributions of formaldehyde, acetaldehyde, and acrolein for the Camp Pendleton bus, Ford F9000 tractor , 250 kW generator, and Humvee respectively. For the Humvee and 250 kW generator, formaldehyde makes the largest carbonyl contribution for all test combinations. For the bus and F900, the distribution between formaldehyde and acetaldehyde is more evenly distributed with the relatively fractions differing depending on the specific test combination. Note that the relative carbonyl emissions showed good reproducibility for a specific engine, irrespective of operating mode or fuel type, especially for the FTP weighted Humvee and the 250 kW generator tests.

4.3.2.3. Gas-Phase Hydrocarbon Species

Additional speciation was performed on the light hydrocarbons. Weighted benzene emissions (Figure 4.27) for the Ford F9000 tractor is provided to demonstrate that other individual species also follow similar emission trends to those compounds already discussed, with Mode 1 dominating the net weighted emissions with mode 8 providing the second most important bin.

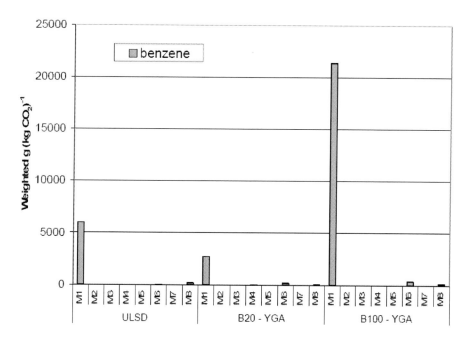

Figure 4.27. F9000 Modal Benzene Emissions Weighted by Mode.

Figure 4.28 presents the weighted FTP benzene and 1,3-butadiene emission rates for the Humvee for ULSD, YGA B20, YGA B100, and JP8. The emissions of 1,3-butadiene are seen to slightly increase in the YGA fuels as compared with ULSD. The 1,3-butadiene emissions are also measured to be higher in JP-8 as compared with the base ULSD fuel, although greater variability in 1,3 butadiene measurements is noted for the JP8 fuel as compared with the YGA fuels and ULSD fuel. No clear trends are noted for benzene across fuels tested for the Humvee.

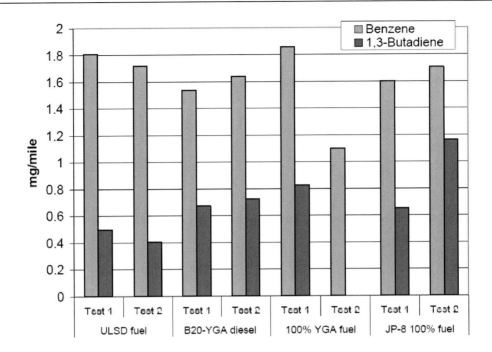

Figure 4.28. Humvee FTP Weighted Benzene and Butadiene Emissions on a Per Mile Basis.

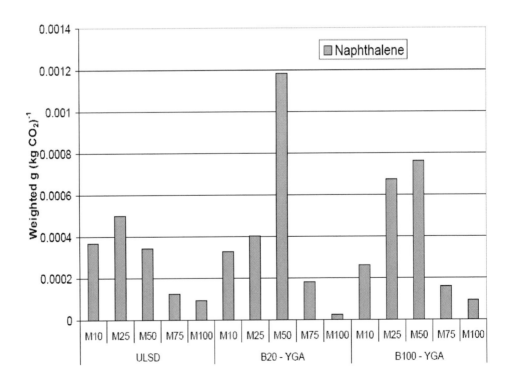

Figure 4.29. 250 kW Generator Modal Naphthalene Emissions Weighted by Mode.

Detailed modal Naphthalene measurements were acquired for the 250 kW generator. Weighted emission factors for this generator are found in Figure 4.29. For the 250 kW generator, naphthalene emissions tend to be lower on a CO_2 basis for the higher load point, although it is expected that the absolute emissions would be greater under these conditions. A peak in the weighted emissions for Naphthalene is noted at 50% power for YGA B20 and YGA B100 for the 250 kW generator. The data for the 250 kW generator show some differences in tests run on different fuels, but the data are limited. Additional measurements of Naphthalene weighted emissions for a subset of modes are provided in Figures 4.30 and 4.31 for the Camp Pendleton bus and Ford F9000 tractor, respectively. For the bus, data are only available for Mode 8. For the F9000, Mode 1 appears to continue to be the most significant mode for weighted mobile source air toxic emissions for YGA B20 following the trends for carbonyl compounds as well as EC and OC. Relatively high naphthalene emissions were also found for the YGA B100 on Mode 6. As these data are relatively limited on the individual vehicles, there are no conclusive fuel trends.

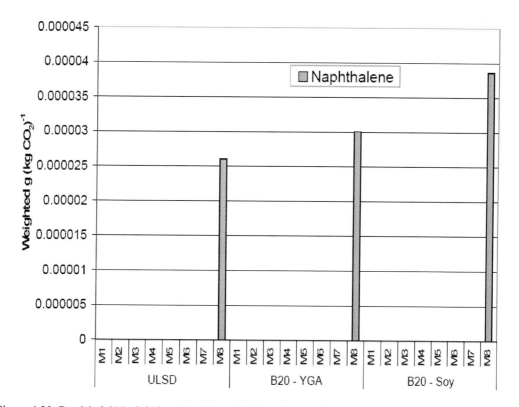

Figure 4.30. Bus Modal Naphthalene Emissions Weighted by Mode.

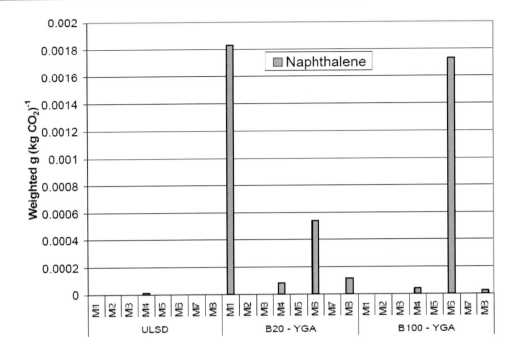

Figure 4.31. F9000 Modal Naphthalene Emissions Weighted by Mode.

4.3.2.4. Statistical Analysis of Fuel Effects for Unregulated Emissions

A detailed statistical analysis was performed to determine the significance of measured fuel effects. Since speciation data is not available for all modes, analyses and comparisons based on weighted values could not be conducted for the species. To provide some understanding of potential trends, a paired t-test was conducted using all the data modes with the vehicle/fuel average data for each mode. Since the analyses are conducted over a wider range of mode representing different operating conditions, a more stringent criteria for statistical significance of $p=0.01$ was applied for statistically significant differences, $p=0.05$ used for marginally statistically significant results. Table 4.5 below summarizes the findings for the three engines. Cells blacked out indicate that insufficient data was collected for analyses of fuel effects, cells highlighted in yellow are those compounds that have a fuel effect (>95% confidence). All white cells indicate that measured fuel effects were **not** statistically significant. An insufficient number of FTP tests were performed to report significant statistical information on the Humvee. Although some differences were found to exist for specific vehicle/fuel combination comparisons, no consistent trends were observed for the data.

Table 4.5. Summary of Emissions Changes and Statistics Relative to ULSD for HAPs

	Engine		EC	OC	TC	Acetaldehyde	Acrolein	Benzene	Butadiene	Formaldehyde	Naphthalene
YGA 20%	Camp	% change	-11.834	4.459	-3.697	21.702				14.704	
		t-test	0.124	0.792	0.158	0.232				0.018	
	Pendleton Bus	% change									
		t-test									
	F9000	% change	5.808	-16.481	-1.979	-36.934	1110.738	538.116		-46.037	547.751
		t-test	0.637	0.178	0.694	0.043	0.195	0.456		0.018	**
	250 kW	% change				44.218	-39.649			36.667	36.767
		t-test				0.239	0.117			0.411	0.819
	Generator	% change									
		t-test									
	Humvee	% change				19.59	24.65	-9.8	55.31	13.28	
		t-test						0.014	0.002		
YGA 100%	F9000	% change	-79.174	-31.298	-64.190	-47.235	235.889	411.028		-53.932	214.207
		t-test	0.006	0.171	0.022	0.264	0.193	0.360		0.271	**
	250 kW	% change				36.300	-6.526			44.306	31.096
		t-test				0.016	0.552			0.015	0.631
	Generator	% change									
		t-test									
	Humvee	% change						-16.13	83.48		
		t-test						0.533			
Soy	Camp	% change	-11.834	4.459	-3.697	21.702	29.551			14.704	48.443
		t-test	0.124	0.792	0.158	0.232	**			0.018	**
B20	Pendleton Bus	% change									
		t-test	0.124	0.792	0.158	0.232				0.018	**

** Only one mode acquired, insufficient data for t-test calculation

99% confidence level – statistically significant - $p \leq 0.05$

95% confidence level – marginally statistically significant - $\leq 0.05 \; p \leq 0.10$

data insufficient for statistical analysis.

5. COST ASSESSMENT

5.1. Cost Reporting

Implementing a biodiesel fueling program represents new additional operational costs over existing petroleum diesel fueling activities. The advantage of implementing a biodiesel program over other potential alternative fuels is that biodiesel can be used in most existing diesel engines without modifications to either the engine or fuel storage system and that the fuel can be dispensed from existing fueling stations. Thus a biodiesel program should have very small startup costs. These benefits are not available for competing alternate fuels such as hydrogen or compressed natural gas. The only direct cost for implementing a biodiesel program is the price difference of the fuel, taking into account the slight decrease in fuel economy with biodiesel, which for most fleets will not be noticeable. As of the preparation date of this Final Report, the average national difference between the commercial price of petroleum diesel and B20 is $0.17 per gallon of fuel. For federal government fleets, the Defense Energy Support Center currently charges a $0.14 per gallon premium for B20.

In terms of indirect environmental costs, NFESC has not been able to identify any costs that would change with biodiesel use. For example, permitting and spill plan requirements for fuel dispensing and storage operations would be the same. The only way that biodiesel use will affect environmental compliance costs is in the area of AFV credits. Since the federal government is mandated to purchase AFVs, satisfying this requirement through the use of biodiesel can minimize this program's compliance costs. Assigning a value to an AFV is, however, very difficult. NFESC is not aware of any value assigned to an AFV credit that has been reported in the literature.

Since the fuel cost is the only identified cost difference between the use of petroleum and biodiesel, this is the only cost information that has been entered into Table 5.1. No attempt has been made in the table to use the Environmental Cost Analysis Methodology (ECAM) developed by the National Defense Center for Environmental Excellence. The reasons the ECAM was not used is that it is not required for incorporating the fuel cost information. Cost information for competing alternate fuels will not be incorporated into Table 5.1 since the purpose of this ESTCP project is to obtain air emissions data for biodiesel, and not to justify its use in place of another alternative fuel.

Table 5.1. Types of Costs by Category

Direct Environmental Activity Process Costs				Indirect Environmental Activity Costs		Other Costs	
Start-Up		Operation & Maintenance					
Activity	$	Activity Fuel Purchase (Price Difference)	$/gal $0.14	Activity	$	Activity	$

5.2. Cost Analysis

Currently many DoDoperated non-tactical diesel powered engines are fueled with low sulfur diesel fuel made from petroleum. This is rapidly changing as new B20 biodiesel programs are rolled out throughout DoD. Tactical engines are fueled with JP-8, a higher sulfur containing fuel also derived from petroleum. Starting in 2006, the EPA has mandated that on-road diesel powered vehicles use ULSD. Since this new mandate will be implemented after the completion of this project, the costs for biodiesel will be not be compared to ULSD. As previously discussed in paragraph 5.1, cost information for competing alternate fuels will not be incorporated into the final reports since the purpose of this ESTCP project is to obtain air emissions data for biodiesel, and not to justify its use in place of another alternate fuel.

The retail price of biodiesel reflects its distribution and manufacturing costs, the profit for the distributors and producers as well as the fact that given the current biodiesel demand/ supply balance, buyers are generally willing to pay a premium over the price of petroleum diesel. The manufacturing costs are primarily driven by the cost of the raw material vegetable oil. At current soybean oil prices, the oil costs a little over one half the retail price of biodiesel.

Generally, the manufacturing cost for making the biodiesel is proprietary information, however, it is believed to be decreasing in recent years as the biodiesel market has developed. This trend is expected to continue for, at least, the next few years.

The costs and differences in cost between petroleum diesel and biodiesel continuously change over the course of time. Petroleum diesel costs are driven by the world cost of crude oil, the oil refining markup and any local supply/demand imbalances. Since biodiesel consumed in the United States is primarily made from virgin soybean oil, its cost is driven by this commodity's price, as well as any local supply/demand imbalances. Currently, the cost of soybean oil for use in the manufacture of biodiesel has been lowered through a $1.00 per gallon direct federal subsidy. The duration and extent of future subsidies is unknown.

In summary, the life-cycle costs for implementing a biodiesel fueling program are totally dependant on the difference in cost between the two fuels. Assuming that the cost for crude oil increases faster than that of soybean oil, the price premium for biodiesel should decrease in the future. This cost difference between petroleum diesel and biodiesel may or may not be important based on any future mandates to use alternate fuels. In this case, the costs for biodiesel must be compared to that of the other alternate fuels.

6. IMPLEMENTATION ISSUES

6.1. Environmental Checklist

Performance of a biodiesel emissions testing or implementation program for diesel-powered vehicles is not expected to require any new environmental permits nor permit changes since no existing pollution control equipment will be modified or removed. This lack of requirements for permitting actions may not be the case for stationary engines. Stationary-source diesel emissions are regulated differently from mobile sources. Generally, they are permitted by state and local air pollution control agencies with unique requirements. Depending on the use of the engine (prime power or standby) and whether or not it is located within a designated air pollution nonattainment area, the switch to biodiesel could potentially trigger a requirement for a permit change. Since each permit is unique, an assessment must be made on a case-by-case basis prior to testing or implementing a biodiesel project on a stationary diesel-powered engine. The project team will work with the owners of stationary diesel engines to be tested to ensure that any required air emissions permit changes are approved prior to initiating testing.

6.2. Other Regulatory Issues

At the beginning of the project CARB and EPA were given a chance to review our project demonstration plan and make suggestions on changes that would provide significant value to their organizations. Both agencies were invited to attend the project's kickoff meeting. At the conclusion of this demonstration project, these same organizations will be given a copy of the ESTCP final report.

In addition to the planned coordination with the environmental regulatory agencies, NFESC also plans to work with the National Biodiesel Board. The purpose of this coordination is to ensure that this ESTCP project does not duplicate other current industry or academic efforts and to assist with the dissemination of project results to interested parties in the biodiesel industry. The participation of other non government organizations will be encouraged and is actively being investigated.

6.3. End-User Issues

The end users of this project will be DoD diesel-powered fleet operators and the DoD diesel fuel suppliers, primarily DLA. Their primary focus with implementing biodiesel fueling programs concern the issues of fuel availability, cost, performance, and environmental regulation. Negative results in any of these areas, will stop an implementation program in its tracks.

Biodiesel may be manufactured from a multitude of agricultural raw materials; however, biodiesel fuel made from virgin soybean oil is the most common type. Although this fuel is overwhelmingly manufactured in the midwestern states, near where the soybeans are grown, it is widely available throughout the continental United States. Significant increases in DoD use of this fuel are not expected to greatly change its supply/demand balance. Unfortunately, at this time, virgin soybean biodiesel costs more than petroleum diesel.

Another potential source of biodiesel is that made from yellow grease. This fuel has an advantage in that it is currently cost competitive with petroleum. As the market for yellow grease is in its early stage of development, this fuel is not widely available. It is generally not used by DoD even though its use have been approved by the Defense Energy Support Center, the primary DoD fuel supplier.

Since B20 biodiesel has an approximately 1.7 % lower fuel energy content per gallon of fuel than petroleum diesel, a small decrease in fuel economy, as measured in miles per gallon, may occur with its use. Fleet operators do not; however, expect any noticeable decrease in engine performance nor any increase in required maintenance. Based on previous biodiesel implementation efforts these performance expectations should be met.

In the area of environmental regulatory compliance, the DoD fleet operator is not expected to face any implementation barriers, just benefits. DoD fleet operators are required to purchase 75 percent of their light duty vehicles capable of operating on alternative fuels.

For every 2,250 gallons of B20 biodiesel fuel used on vehicles weighing more that 8,500 pounds, one AFV credit is allowed. Since the use of biodiesel will not require any infrastructure upgrades, its widespread implementation is an easy way to earn AFV credits and thus avoid other costly options such as the use of natural gas powered vehicles.

In addition to these current regulatory issues, it is expected that some form of air pollution reductions will be mandated for existing diesel engines particularly in the area of PM emissions. Since B20 biodiesel has been reported to reduce PM emissions when compared with current low sulfur Diesel Fuel No. 2 (see Table 2.1), the DoD fleet operator may avoid other more costly retrofit requirements.

To ensure that project results are transitioned to DoD fleet operators, the transition plan for this project will involve publicizing the test results in various forms that are readily available to DoD and regulatory decision makers. This publicizing effort will include providing a copy of the final report to the National Biodiesel Board as well as making a presentation of the results at the National Biodiesel Board annual brainstorming session. Since the DLA controls the accepted DoD buy list, NFESC will work with that organization to ensure that potential biodiesel users fully understand its potential benefits.

To reach potential DoD interested parties, NFESC has presented project results at the annual Joint Service Environmental Management Conference as well as a regional Air and Waste Management Association and Federal Laboratory Consortium Conference. Currently, NFESC produces a number of products supporting the transition of technologies to the Navy and other DoD customers. These products include generating environmental quality initiative fact sheets, Currents Magazine articles, Pollution Prevention Technical Library data sheets, pocket cards, user data packages, technical reports, technology implementation plans and a point of contact (POC) list of potential customers. Project results will also be posted on the Defense Environmental Network & Information Exchange (DENIX). Potential non-DoD interests will be informed of the results of the project by submitting articles for publication in applicable trade publications and technical journals, as well as using the National Biodiesel Board to disseminate information.

7. REFERENCES

7.1 EPA Draft Technical Report EPA 420-P-02-001 "A Comprehensive Analysis of Biodiesel Impacts on Exhaust Emissions"

7.2 National Renewable Energy Laboratory Report NREL/SR-510-31465 "NOx Solutions for Biodiesel", RL McCormick, JR Alvarez, MS Graboski

7.3. Siegl, W.O.; Richert, J.F.O.; Jensen, T.E.; Schuetzle, D.; Swarin, S.J.; Loo, J.F.; Prostak, A.; Nagy, D.; and Schlenker, A.M. (1993) Improved Emissions Speciation Methodology for Phase II of the Auto/Oil Air Quality Improvement Research Program – Hydrocarbons and Oxygenates. SAE Technical Paper 930142. Society of Automotive Engineers, Warrendale, PA.

7.4. Shah et al., 2005, Environmental Science & Technology, 39, 5976-5984.

7.5. Birch and Cary, 1996, Aerosol Sci. Technol. 25, 221-241.

7.6. Shah et al., 2004, Environmental Science & Technology, 38:9, 2544-2550.

7.7. Cocker III, D. R., Shah, S., Johnson, K., Miller, J. W., Norbeck, J., Development and Application of a Mobile Laboratory for Measuring Emissions from Diesel Engines. I Regulated Gaseous Emissions, Environ. Sci. Technol.,**2004**, 38, 2182-2189.

7.8. Code of Federal Regulations Title 40, Part 86, Subpart B.

7.9. http://www.dieselnet.com/standards/cycles/avl_8mode.html

7.10. US Environmental Protection Agency, 40 Code of Federal Regulations (CFR) Part 89 – Control of Emissions from New and In-use Nonroad Compression Ignition Engines, June 28, **2002.**

7.11. International Standards Organization, IS0 8178-4, Reciprocating internal combustion engines - Exhaust emission measurement -Part 4: Test cycles for different engine applications First edition 1996-08-15.

7.12. Sharp, C.A., (1997) Biodiesel Effects on Diesel Engine Exhaust Emissions, Biodiesel Workshop, April 15.

7.13. Sharp, C.A., (1998) Characterization of Biodiesel Exhaust Emissions for EPA 211(b), Final Report for the National Biodiesel Board.

7.14. Sharp, C.A., (1998) The Effects of Biodiesel on Diesel Engine Exhaust Emissions and Performance, Biodiesel Environmental Workshop.

7.15. Graboski, M.S.; Ross, J.D.; McCormick, R.L., (1996) Transient Emissions from No. 2 Diesel and Biodiesel Blends in a DDC Series 60 Engine. SAE Technical Paper No. 961166.

7.16. Smith, J.A.; Endicott, D.L.; and Graze, R.R, (1998) Biodiesel Engine Performance and Emissions Testing. Final Report prepared for the National Biodiesel Board, May.

7.17. Spataru, A. and Romig, C., (1995) Emissions and Engine Performance of Soya and Canola methyl esters Blended with ARB #2 Diesel Fuels, Final Report to Saskatchewan Canola Development Commission.

7.18. McDonald, J.F.; Purcell, D.L.; McClure, B.T.; and Kittelson, D.B, (1995) Emissions Characteristics of Soy Methyl Ester Fuels in an IDI Compression Ignition Engine, SAE Technical Paper No. 950400.

7.19. Clark, N.N.; Atkinson, C.M.; Thompson, G.J.; and Nine, R.D., (1999) Transient Emissions Comparisons of Alternative Compression Ignition Fuels. Submitted to 1999 SAE Congress.

7.20. Starr, M.E., (1997) Influence on Transient Emissions at Various Injection Timings, using Cetane Improvers, Biodiesel, and Low Aromatic Fuels, SAE Technical Paper No. 972904.

7.21. Choi, C.Y.; Bower, G.R.; Reitz, R.D. Effects of Biodiesel Blended Fuels and Multiple Injections on D.I. Diesel Engines; SAE Technical Paper No. 970218; Society of Automotive Engineers: Warrendale, PA, 1997.

7.22. Akasaka, Y.; Suzuki, T.; Sakurai, Y. Exhaust Emissions of a DI Diesel Engine Fueled with Blends of Biodiesel and Low Sulfur Diesel Fuel; SAE Technical Paper No. 972998; Society of Automotive Engineers: Warrendale, PA, 1997.

7.23. Durbin, T. D., J. R. Collins, J. M. Norbeck and M. R. Smith. 2000b. The Effects of Biodiesel, Biodiesel Blends, and a Synthetic Diesel on Emissions from Light Heavy-Duty Diesel Vehicles. Environ. Sci. & Technol., 34, 349.

7.24. Durbin, T. D. and J. M. Norbeck. 2002. The Effects of Biodiesel Blends and ARCO ECDiesel on Emissions from Light Heavy-Duty Diesel Vehicles. Environ. Sci. & Technol., vol. 36, 1686-1691.

8. POINTS OF CONTACT

Table 8.1. Points of Contact

POINT OF CONTACT Name	ORGANIZATION Name Address	Phone/Fax/Email	Role in Project
Bruce Holden	NFESC 1100 23RD Avenue Port Hueneme, CA. 93043	(805) 982-6050 (voice) (805) 982-1409 (fax) bruce.holden@navy.mil	Principal Investigator
Dr. Norman Helgeson	NFESC 1100 23RD Avenue Port Hueneme, CA. 93043	(805) 982-1335 (voice) (805) 982-4832 (fax) norman.helgeson@navy.mil	Quality Assurance Officer
Jason Jack	U.S. Army Aberdeen Test Center CSTE-DTC-AT-SL-S 400 Colleran Road APG, MD. 21005-5059	(410) 278-4045(voice) (410) 278-1589(fax) Jason.jack@atc.army.mil	Environmental Scientist
Dr. Wayne Miller	University of California, Riverside CE-CERT 1084 Columbia Ave. Riverside, CA. 92507	(909) 781-5579 (voice) (909) 781-5590 (fax) wayne@cert.ucr.edu	Director Emissions & Fuels Research Laboratory
Dr. Tom Durbin	University of California, Riverside CE-CERT 1084 Columbia Ave. Riverside, CA. 92507	(909) 781-5794 (voice) (909) 781-5590 (fax) durbin@cert.ucr.edu	Associate Research Engineer
Bob Hayes	ReFUEL Lab - National Renewable Energy Laboratory 1617 Cole Blvd. Golden, CO 80401	(303) 275-3143 (voice) (303) 275-3147 (fax) Bob_Hayes@nrel.gov	Senior Engineer

APPENDIX A.
PHOTOGRAPHS OF TEST ENGINES AND EQUIPMENT

Military Humvee in the UCR Vehicle Emissions Research Laboratory

**Ford F350 Pick-up Truck in the UCR
Vehicle Emissions Research Laboratory**

Ford F9000 Truck Pulling the CE-CERT Mobile Emissions Laboratory

Camp Pendleton Bus at Heavy-Duty Chassis Facility

Ford F700 Truck at Heavy-Duty Chassis Facility

Military Generator Testing with CE-CERT Mobile Emissions Laboratory

Aircraft Tow Tractor Testing with ATC Rover System

Cheyenne Mountain Air Force Base Bus Being Tested at the National Renewable Energy Laboratory in Denver, CO

Hyster Forklift Being Tested at ATC

APPENDIX B.
DATA QUALITY ASSURANCE/QUALITY CONTROL PLAN

B.1. Purpose and Scope of the Plan

The purpose of the Quality Assurance (QA) plan is to ensure that the data collected during this demonstration project is of sufficient quality to fulfill the project objectives.

B.2. Quality Assurance Responsibilities

The primary responsibility for Quality Assurance belongs to each project performer. However, Dr. Norman Helgeson, an NFESF engineer will serve as the project's Quality Assurance Officer. A NFESC senior engineer, not directly involved in the project, will provide peer review of the final report.

B.3. Data Quality Parameters

Gaseous air emissions will be collected using electronic instruments with the results directly transferred to a computer. For the ATC tests utilizing the EPA's ROVER system, two NOx gas analyzers are employed. Data averaging and integration functions will be performed within the computer. Emission results for each of the gaseous criteria pollutants will be transferred to and stored in a laboratory notebook maintained for this project. Since the measured quantities are time sensitive, sample storage and re-testing is not possible. Weight records for PM collections will likewise be recorded in the project's laboratory notebook. At each test condition, a minimum of two tests will be performed. If the results from any test differ more than the expected standard deviation from an equivalent previous test, an investigation will be performed to try and identify any test equipment or other problems.

The chosen test cycles must satisfy two needs. They must closely represent the actual duty cycle placed on the engine while also being a cycle commonly reported in the literature so that testing results can be compared with those from other investigations. Since these two needs are not equivalent, the chosen test cycle(s) represents a compromise.

B.4. Calibration Procedures, Quality Control Checks, and Corrective Action

All the project's air emissions measuring instruments including the gaseous sampling equipment and the analytical balances used for weighing the PM samples are maintained under calibration programs run by the respective testing organizations. Quality inspections that are mandated by the CFR, such as propane balances within the CVS system, are run weekly and corrective actions taken as needed. Prior to each test performed by ATC using the ROVER, gas calibrations and pressure checks are completed and recorded to verify accurate installation of the system and accuracy of its components. In addition to the test instruments calibration, the dynamometers and electric load banks used to set the engine load points will be checked prior to each use. As previously stated, testing at each testing condition will be repeated a minimum of two times.

Most test data will be collected electronically with a computer integrating and averaging the data. Other data will be directly recorded in a laboratory notebook. Data will be reported in tabular format using standard units so that it can easily be compared with results from other investigations

B.5. Demonstration Procedures

The test of each engine will begin with either the vehicle being transported to the testing laboratory or field-testing instruments brought to the engine to be tested. It is expected, that in total, testing can be completed on each engine within a five week time period. Problems with the testing equipment or the engine will be addressed on a case-by-case basis.

Since multiple fuels will be tested on each engine, the testing organizations will be required to mix and store fuels as well as performing multiple filling and draining of fuel systems. All fuels for the project will be distributed out of a central storage location. Commercial fuel transporters will be used.

B.6. Calculation of Data Quality Indicators

Data quality will be determined by comparing test results from identical test conditions, by comparing the results with those from previous investigations, and finally by comparing the results with the EPA models (see Reference 7.1). In addition, UCR maintains quality control charting of the calibration gases for both the light- and heavy-duty emissions laboratories. Such charts allow for the taking of corrective actions when the system is out of control as defined in the quality manuals. For example, five points in a row below the mean value will require a corrective action, even though the measured value is within the upper and lower control limits.

B.7. Performance and System Audits

To ensure that the collected data represents the actual system conditions, an NFESC engineer/scientist, Dr. Norman Helgeson will independently audit the emissions measuring techniques. The purpose of this audit is to ensure that the measuring method accurately reflects the actual demonstration conditions. The audit will consist of laboratory and field measurement reviews, review of instrument calibration certifications and a specific review of the air emissions measuring procedure. It is expected that one audit will be completed.

During 2002, the EPA conducted a quality audit of the light- and heavy-duty laboratories at UCR and both were found to conform within the expected specifications for QA/QC. Additionally, the light-duty laboratory participates in an annual cross-laboratory round robin wherein a vehicle is tested at up to 25 laboratories. In each of the last two years, the UCR laboratory has measured values well within the variation of all the participating laboratories. In addition, the heavy-duty laboratory was verified by comparing the values obtained in the CARB heavy-duty laboratory with those of the UCR laboratory. Agreement was within the values found in a recent publication for a round robin study.

B.8. Quality Assurance Reports

It is expected that one audit will be completed. Any significant results from the Performance and System Audit will be incorporated into the final project reports. Quality assurance status reports will not be prepared. This decision is based on the testing organizations long history of successfully performing similar testing operations.

B.9. ISO 14001

NFESC, the principle organization responsible for this demonstration, is not ISO 14001 certified. UCR likewise is not ISO 14001 certified, however, it has a Quality Plan prepared for its organization, and the light-duty and heavy-duty laboratories have a number of SOPs and quality control practices that are followed for all test programs. ATC is in the process of obtaining ISO 14001 certification. At the time of the preparation of this demonstration plan it is not known when the certification will be obtained.

B.10. Data Format

Test data will be acquired both electronically and on paper. At UCR, gaseous emissions are measured continuously from a number of instruments and the results stored on a computer disk within the light- and heavy-duty laboratories. These data are stored along with frequent calibrations of the instruments using calibration gases that are generated in a gas divider. Other data for the integrated PM mass or for HAPs are measured off-line and then integrated into the data page where the averages for the continuous gaseous measurements.

ATC ROVER test data is electronically generated in spreadsheet format. A data point is generated for each second of the test. This raw data will be submitted along with a separate ATC test record to the principal investigator for review.

B.11. Data Storage and Archiving Procedure

All correspondence, documentation, raw data, and records generated as a result of this demonstration project will be maintained on site by the Principal Investigator for a period of one year after projection completion. The information will be collected in its originally generated form (i.e. paper or electronic). The ESTCP Final Report, in paper and electronic versions, will be maintained by the NFESC Technical Information Center. The POC for the Technical Information Center is Bryan Thompson at (805) 982-1124

In: Biodiesel Fuels Reexamined
Editor: Bryce A. Kohler

ISBN: 978-1-60876-140-1
© 2011 Nova Science Publishers, Inc.

Chapter 2

EFFECTS OF BIODIESEL BLENDS ON VEHICLE EMISSIONS: FISCAL YEAR 2006 ANNUAL OPERATING PLAN MILESTONE 10.4[*]

R.L. McCormick, A. Williams, J. Ireland, M. Brimhall and R.R. Hayes

EXECUTIVE SUMMARY

Biodiesel is a fuel-blending component produced from vegetable oils, animal fats, or waste grease by reaction with methanol or ethanol to produce methyl or ethyl esters. Pure biodiesel contains approximately 10 weight percent oxygen. It is typically blended with petroleum diesel at levels up to 20% (B20). The presence of oxygen in the fuel leads to a reduction in emissions of hydrocarbons (HC) and toxic compounds, carbon monoxide (CO), and particulate matter (PM) when biodiesel blends are burned in diesel engines. These reductions are robust and have been observed in numerous engine and vehicle testing studies. Engine dynamometer studies reviewed in a 2002 report from EPA show a 2% increase in oxides of nitrogen (NO_x) emissions for B20. This perceived small increase in NO_x is leading some state regulatory agencies to consider banning the use of biodiesel. Therefore, the issue of NO_x emissions is potentially a significant barrier to expansion of biodiesel markets.

The objective of this study was to determine if testing entire vehicles, vs. just the engines, on a heavy-duty chassis dynamometer provides a better, more realistic measurement of the impact of B20 on regulated pollutant emissions. This report also documents completion of the National Renewable Energy Laboratory's Fiscal Year 2006 Annual Operating Plan Milestone 10.4. This milestone supports the U.S. Department of Energy, Fuels Technologies Program Multiyear Program Plan Goal of identifying fuels that can displace 5% of petroleum diesel by 2010.

---

[*] This is an edited, reformatted and augmented version of a National Renewable Energy Laboratory's publication, dated October 2006.

We reviewed more recently published engine testing studies (Table 3) and found an average change in NO_x for all recent B20 studies of -0.6%±2.0% (95% confidence intervals are used throughout this report). Restricting the average to recent studies of B20 with soy biodiesel yields an average NO_x impact of 0.1%±2.7%. The EPA review also includes summary of a smaller vehicle testing dataset that shows no significant impact of biodiesel on NO_x. We reviewed several recently published vehicle (chassis) testing studies (Tables 4 and 5) and found an average change in NO_x of 1.2%±2.9% for B20 from soy-derived biodiesel. In addition, we reviewed three portable emissions measurement system (PEMS) studies that do not find NO_x to increase.

Eight heavy-duty diesel vehicles were tested, including three transit buses, two school buses, two Class 8 trucks, and one motor coach. Four met the 1998 heavy-duty emissions requirement of 4 g/bhp-h NO_x and four met the 2004 limit of 2.5 g/bhp-h NO_x+HC. Driving cycles that simulate both urban and freeway driving were employed. Each vehicle was tested on a petroleum-derived diesel fuel and on a 20 volume percent blend of that fuel with soy-derived biodiesel. On average B20 caused PM and CO emissions to be reduced by 16% to 17% and HC emissions to be reduced by 12% relative to petroleum diesel. Emissions of these three pollutants nearly always went down, the exception being a vehicle equipped with a diesel particle filter that showed very low emissions of PM, CO, and HC; and there was no significant change in emissions for blending of B20. The NO_x impact of B20 varied with engine/vehicle technology and test cycle ranging from -5.8% to +6.2%. A preliminary examination of real-time NO_x emission data did not reveal any consistent reason for the wide range. On average NO_x emissions did not change (0.6%±1.8%). If the results of this study are combined with the soy B20 chassis results from Tables 4 and 5 (recently published studies), the average change in NO_x is 0.9%±1.5%, based on data for 15 vehicles.

Based on the studies reviewed and new data reported here, there does not appear to be a discrepancy between engine and chassis testing studies for the effect of B20 on NO_x emissions. Individual engines may show NO_x increasing or decreasing, but on average there appears to be no net effect, or at most a very small effect on the order of ±0.5%. The small apparent increase in NO_x reported for engine-testing results in EPA's 2002 review occurred because the dataset was not adequately representative of on-highway engines. In particular, nearly half of the NO_x observations included in the review were for engines from a single manufacturer (DDC). Newer engine and chassis studies, which on average show no B20 effect on NO_x, are not representative samples either. However, considering all of the data available, we conclude that B20 has no net impact on NO_x.

ACRONYMS AND ABBREVIATIONS

Bxx	biodiesel blend containing xx% biodiesel
CARB	California Air Resources Board
CFR	Code of Federal Regulations
CILCC	combined international local and commuter cycle
CO	carbon monoxide
CO_2	carbon dioxide
CSHVC	city-suburban heavy-vehicle cycle

CVS	constant volume sampling
DC	direct current
DDC	Detroit Diesel Corporation
DOC	diesel oxidation catalyst
DPF	diesel particle filter
EGR	exhaust gas recirculation
EPA	U.S. Environmental Protection Agency
FTP	federal test procedure
g/bhp-h	grams per brake horsepower-hour
g/cc	grams per cubic centimeter
HC	hydrocarbon
hp	horsepower
IDI	indirect injection
L	liter
lb	pounds mass
LSD	low-sulfur diesel (<500 ppm S)
mph	miles per hour
NO	nitric oxide
NO_x	oxides of nitrogen
NREL	National Renewable Energy Laboratory
O_2	oxygen
PEMS	portable emissions measurement system
PM	particulate matter
ppm	parts per million
REE	rapeseed ethyl ester
ReFUEL	Renewable Fuels and Lubricants Laboratory
RME	rapeseed methyl ester
RTD	Regional Transportation District
RUCSBC	Rowan University Composite School Bus Cycle
SME	soy methyl ester
TDI	turbocharged-direct injected
THC	total hydrocarbon (the same as HC in this study)
TxLED	Texas low emissions diesel
UDDS	urban dynamometer driving schedule
ULSD	ultra-low-sulfur diesel (<15 ppm S)
WVU	West Virginia University

INTRODUCTION

Biodiesel is a fuel-blending component produced from vegetable oils, animal fats, or waste grease by reaction with methanol or ethanol to produce methyl or ethyl esters (transesterification). In the United States, essentially all biodiesel is fatty acid methyl esters. Biodiesel production was 75 million gallons in 2005 and is expected to grow rapidly with market size, reaching 300 million gallons in 2006 [1]. Roughly 90% of the biodiesel produced in the United States today is made from soybean oil. An assessment of the resource available to produce biodiesel indicates that today there is adequate feedstock available to produce more than 1.7 billion gallons per year [2]. Life cycle analysis shows that for soy-derived biodiesel the energy available in the biodiesel product is more than three times the fossil energy used in its production [3].

Pure biodiesel contains approximately 10 weight percent oxygen. It is typically used as a blend with petroleum diesel at levels up to 20% (B20). The presence of oxygen in the fuel leads to a reduction in emissions of hydrocarbons (HC) and toxic compounds, carbon monoxide (CO), and particulate matter (PM) when biodiesel blends are burned in diesel engines [4]. These reductions are robust and have been observed in numerous engine and vehicle testing studies. Engine dynamometer studies conducted mainly in the 1990s have shown a small increase (2%) in oxides of nitrogen (NO_x) emissions for B20, and most studies of the impact of biodiesel on pollutant emissions have employed engine dynamometer tests. This perceived small increase in NO_x is leading some state regulatory agencies to consider banning the use of biodiesel. Therefore the issue of NO_x emissions is potentially a significant barrier to expansion of biodiesel markets. However, a 2% increase in NO_x is only slightly greater than the measurement repeatability of many heavy-duty test labs. Additionally, several engine testing studies have found no increase, or even a decrease in emissions of NO_x.

The objective of this study was to determine if testing entire vehicles on a heavy-duty chassis dynamometer provides a better, more realistic measurement of the impact of B20 on regulated pollutant emissions.

BACKGROUND

Engine Dynamometer Studies

A number of studies have examined the emission impacts of biodiesel in 4-stroke, electronically controlled, turbocharged, direct injected diesel engines [5, 6, 7, 8, 9, 10, 11, 12, 13, 14, 15]. These studies as well as others using 2-stroke, indirect injection (IDI), and naturally aspirated engines have been reviewed by the U.S. Environmental Protection Agency (EPA) [16] and statistical analysis indicated the average emission changes for B20 shown in Table 1. Figure 1, taken from the EPA report, shows the overall trends with biodiesel blending level for all four regulated pollutants. The small increase in NO_x emissions for B20 listed in Table 1 is notable because a 2% change in NO_x is only slightly greater than the test repeatability (coefficient of variation) for NO_x measurements at the best test laboratories.

Table 1. Average Change in Emissions for B20 as Estimated from Published Engine Dynamometer Data in the EPA Study [16]

Pollutant	Percent Change
HC	-21.1
CO	-11.0
NO_x	+2.0
PM	-10.1

The results are derived from published data on 43 different engines of varying model year. These are grouped by emission standard or technology in Table 2. The dataset is dominated by 26 engines in the 1991 to 1993 group (5 g/bhp-h NO$_x$ and 0.25 g/bhp-h PM) and the 1994 to 1997 group (5 g/bhp-h NO$_x$ and 0.1 g/bhp-h PM). Fifteen out of the group of 26 engines were from a single manufacturer, Detroit Diesel Corporation (DDC). Fully 64% of the 546 NO$_x$ observations for this model year range are for DDC engines, and these observations make up 44% of the total NO$_x$ observations. Because of the high concentration of engines from a single manufacturer and in a limited range of model years, this group of engines cannot be considered to be representative of on-highway engines in the United States. Notably the dataset includes only two engines certified at 4 g/bhp-h NO$_x$ and no engines in technology group B certified at 2.5 g/bhp-h NO$_x$+HC. This later group typically employs exhaust gas recirculation (EGR) to obtain lower NO$_x$ levels. The majority of the engines were on-highway heavy-duty engines and were tested over the heavy-duty federal test procedure (FTP) or multimode steady-state cycles.

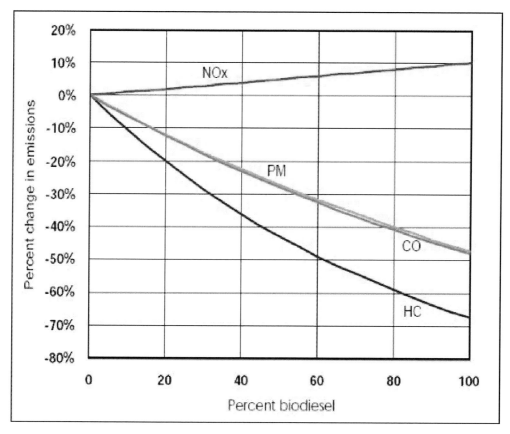

Figure 1. Trends in percentage change in pollutant emissions with biodiesel content as estimated from published engine dynamometer data in the EPA study [16].

**Table 2. Number of Engines and NO$_x$ Emissions
Observations for the Data Reviewed by EPA [16]**

Standards group	Model years	HD highway engines	NOx observations
B	2002-2006	0	0
C	1998-2001	2	14 (2)[a]
D	1994-1997	10	152 (19)
E	1991-1993	16	394 (50)
F	1990	3	87 (11)
G	1988-1989	8	112 (14)
H	1984-1987	2	16 (2)
I	-1983	2	10 (1)

[a] Values in parentheses are percent of total observations.

Figures 2 and 3 show the NO$_x$ and PM emission curves, respectively, as a function of biodiesel blend content, along with the individual data points. Examining the B20 results, a wide range of -60% to +5% is observed for PM; however, PM emissions increased in only one test. A wide range is also observed for NO$_x$ emissions with percentage change ranging from roughly -7% to +7%.

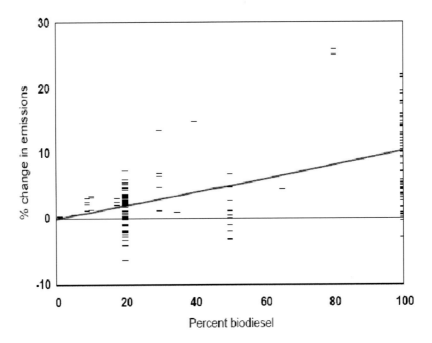

Figure 2. Percent change in NO$_x$ emissions for the engine dynamometer data reviewed by EPA [16].

Additional engine testing studies have been published since the release of the EPA review. McGill and coworkers tested two heavy-duty engines (4- to 5-g/bhp-h NO emission range) [17]. Additional details for a Euro 2 Volvo 9.6-L engine are given in reference 18. NO_x emissions were unchanged for a 5-g/bhp-h NO_x Navistar 7.3-L engine (B100) but increased for a Euro 2 Volvo engine on B30. Frank and coworkers tested a 4 g/bhp-h NO_x International DT466 and observed NO_x to decrease significantly for B20 when configured with DOC, no change when configured with DPF, and to increase slightly when configured with EGR and DPF [19]. A second study tested two 4-g/bhp-h NO Cummins engines and found statistically significant reductions in NO_x for B20 in some tests, and small increases in NO_x in others [20]. Notably, the NO_x reductions were observed for biodiesel from more saturated feedstocks, in agreement with previous studies [4]. Researchers at Penn State University tested a 4-g/bhp-h NO_x Cummins engine and observed a 3% reduction in NO_x for a low sulfur base fuel and no change for an ultra-low sulfur base fuel [21]. Results for two engines meeting the 2.5-g/bhp-h NO_x+HC level have also been reported, with NO_x found to increase by 3 to 6% [22]. Environment Canada has reported testing of a 1998 Caterpillar 3126E with no change in NO_x [23].

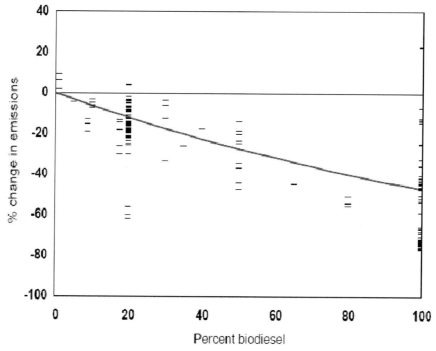

Figure 3. Percent change in PM emissions for the engine dynamometer data reviewed by EPA [16].

The percent change in emissions for these studies relative to the base petroleum fuel is listed in Table 3. The average change in NO_x for all B20 studies reported in this table is -0.6%±2.0% (95% confidence interval). Restricting the average to studies of B20 with soy biodiesel only yields an average NO_x impact of 0.1%±2.7%. Clearly recent engine testing studies continue to see NO_x emission results that vary widely and appear to depend upon engine manufacturer or engine design. The average PM emission change for all B20 studies reported in the table is -14.1%, excluding the two DPF points the PM reduction was 16.4%.

**Table 3. Summary of Percent Change
in Emissions for Recent Engine Dynamometer
Studies of Biodiesel**

Reference	Engine	Cycle	%Biodiesel	NO_x	HC	CO	PM
17	Navistar 7.3-L (5 g/bhp-h NO_x)	AVL 8 Mode	100 (RME)	≈0	--	--	≈-20%
18	Volvo 9.6-L (Euro 2)	ECE R49	30 (RME)	1.7	0	-9.4	-24
19	International DT466 (4 g/bhp-h NO_x	Hot FTP	20 (SME)	-10.3	-20	-38	-2.9
	-with DPF		20 (SME)	0	≈0	≈0	≈0
	-with EGR and DPF		20 (SME)	1.8	≈0	≈0	≈0
20	Cummins 8.3-L (4 g/bhp-h NO_x Mech)	Hot FTP	20 (SME)	1.1	-12	-25	-31
			20 (Waste Grease)	0.3	-7.0	-25	-20
			20 (Animal Fat)	-1.5	-13	-17	-22
	Cummins 8.3-L (4 g/bhp-h NO_x Elec)	Hot FTP	20 (SME)	1.7	-21	-28	-17
			20 (Waste Grease)	-4.5	-25	-31	-14
Reference	Engine	Cycle	%Biodiesel	NO_x	HC	CO	PM
			20 (Animal Fat)	-2.9	-30	-25	-7.8
21	Cummins 5.9-L (4 g/bhp-h NO_x)	AVL 8 Mode	20 (SME, 325 ppm S Base)	0	--	--	-27
			20 (SME, 15 ppm S Base)	-3	--	--	-6
22	Cummins 5.9-L (2.5 g/bhp-h NO_x)	Hot FTP	20 (SME)	3.6	-4.2	-10.5	-22
	DDC S60 (4 g/bhp h NO_x)	Hot FTP	20 (SME)	6.0	0	0	-26
23	Caterpillar 3126E (4 g/bhp-h NO_x)	Hot FTP	20 (SME)	0	-16	-6.7	-1.1

Vehicle Testing Studies

EPA's review [16] also included a summary of chassis dynamometer vehicle testing studies. The studies reviewed included data for three transit buses and eight pickup trucks and the data for percent change in NO_x emissions are shown in Figure 4. While a fitted trend line shows a negative slope (i.e., NO_x emissions being reduced as biodiesel is blended with diesel fuel), the slope of this line is not significantly different from zero (p 0.5), thus for these vehicles blending of biodiesel had no impact on NO_x.

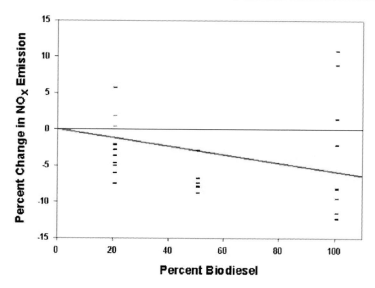

Figure 4. Summary of NO_x impact of biodiesel blending from chassis dynamometer studies reviewed by EPA [16].

A number of chassis studies have been published since the publication of EPA's review, or were not included in that review. Careful review of the data in these publications reveals large variation in data quality, with some studies exhibiting extremely poor repeatability (likely because an inadequate number of control tests were conducted), or basing conclusions on only one or two replicate tests. Application of strict quality criteria, and rejection of studies that do not meet them, is required for any discussion of chassis testing data.

Data from many of these studies are shown in Table 4. Clark and Lyons reported testing eight Class 8 tractors, ranging in model year from 1989 to 1994 on conventional diesel and B35 [24]. NO_x emissions over the WVU 5-peak cycle decreased for two vehicles, increased for five vehicles and were unchanged for one vehicle. A second report from the same research group includes testing of additional vehicles and suggests that NO_x emissions decrease for vehicles with early 1990s DDC engines, but increase for late 1980s Cummins engines [25]. However, no individual vehicle results are reported in this study, and it is not clear if intake air humidity was controlled or measured in these tests, hence they are not considered further. Petersen and coworkers report test results for a Dodge pickup equipped with a 1994 Cummins B5.9 [26] (it is not clear if these results were included in EPA's review). Biodiesel from several different sources was tested and reductions in NO_x were observed in all cases for B20 blends. Three replicate tests were run only for the B20 produced from rapeseed ethyl ester (REE), which produced a 3.1% decrease in NO_x. Durbin and coworkers present light-duty FTP results for testing B20 in seven light heavy-duty diesel vehicles with model year ranging from 1983 to 1993 [27]. However, this study is based on only two replicate runs with inadequate controls and thus will not be considered further. Finnish researchers reported testing an Audi turbocharged direct injected (TDI) vehicle on the light-duty FTP with B30 and adequate replication and controls. They observed no significant change in NO_x [18]. Environment Canada tested three heavy-duty trucks on B20 and showed that NO_x could go up or down depending on engine design [28]. Most recently Holden and coworkers have presented a significant study of on- and off-highway nontactical vehicles used at military bases [29]. All vehicles were tested multiple times with California Air Resources Board

(CARB) diesel base fuel tests at regular intervals. For the study results taken as a whole, and taking into account only changes in emissions that are significant at 90% confidence or better, there was no significant impact of B20 on NO_x. Results from this study are shown in Table 5. The results in Tables 4 and 5 for soy-derived B20 show an average change in NO_x of 1.2%±2.9% (95% confidence interval). Both the chassis studies reviewed by EPA and the more recent studies described here are showing no significant impact of B20 on NO_x.

Table 4. Summary of Percent Change in Emissions from Recent Vehicle Testing Studies of Biodiesel

Reference	Engine	Cycle	%Biodiesel	NO_x	HC	CO	PM
26	1994 Cummins ISB	UDDS	20 (REE)	-3.1	-36	-37	-12
18	Audi TDI	FTP75	30 (RME)	0	-13	5.5	-22
28	2003 Cummins ISM	UDDS	20 (SME)	-3.1	-8.2	-16	-20
		WVU 5 Pk	20 (SME)	-2.5	-23	-19	12
	2004 MBE4000	UDDS	20 (SME)	14	-23	-19	-20
	1999 Caterpillar C12	UDDS	20 (SME)	3	-21	-17	-27
		WVU 5 Pk	20 (SME)	1.5	-21	-7.6	-2.9

Another approach to vehicle testing is the use of portable emissions measurement systems (PEMS). These are systems that reside on board the vehicle during normal operation or operation on a test track, and measure concentrations of pollutants in the raw exhaust. EPA will use PEMS to assess in-use compliance of heavy-duty vehicles with emission standards beginning in the 2007 model year [30]. Several measurements of the impact of B20 on emissions have been conducted using PEMS. Frey and Kim reported on testing 12 Department of Transportation dump trucks during their normal operation in North Carolina [31]. These included engines from four manufacturers and model years from 1998 to 2004. The study measured HC, CO, and NO emissions, but not NO_x emissions, such that emissions of NO2 are not included. On average, emissions of all measured pollutants decreased, with both NO and PM emissions declining by 10%. This strongly implies that NO_x emissions also decreased because it has been shown that the cycle average NO2/NO_x ratio does not change for B20 [23]. Researchers at Rowan University tested three school buses on a test track with emissions measurement by PEMS using a highly aggressive driving cycle developed from bus activity data [32]. The school buses included a 1996 Cummins B-series engine and two 1997 International engines. NO_x emissions went up slightly for two buses and down slightly for the third. Researchers at the Texas Transportation Institute tested five school buses selected to be representative of the school bus fleet in Texas, on a test track using cycles derived from bus activity data and emissions measurement by PEMS [33]. All were equipped with in-line, 6-cylinder International engines ranging in model year from 1987 to 2004. The tests used Texas low emissions diesel (TxLED) as the base fuel (a low aromatic, high cetane number fuel), biodiesel derived from soy, and a market average biodiesel blend (compositional details were not specified). Changes in NO_x emissions were small and not statistically significant.

Table 5. Summary of Percent Change in Emission Results from U.S. Navy Study of B20 Emission Impacts [29]

Vehicle	Cycle		YGA - B20				Soy - B20				YGB - B20			
			HC	CO	NOx	PM	HC	CO	NOx	PM	HC	CO	NOx	PM
F350	FTP	% change	-19%	-8%	0%	-15%								
		p-value	0.15	0.03	0.65	0.21								
	US06	% change	-18%	-3%	24%	-13%								
		p-value	0.31	0.53	0.32	NA								
Model A2 Humvee	FTP	% change	93%	17%	1%	0%	113%	26%	-1%	-9%				
		p-value	0.17	0.06	0.48	0.99	0.06	0.04	0.61	0.05				
	US06	% change	2%	19%	2%	-44%	3%	44%	-1%	-57%				
		p-value	0.91	0.01	0.01	0.32	0.88	0.01	0.09	0.23				
F700	AVL 8-mode	% change					4.8%	2.7%	3.2%	5.6%	0.4%	-2.1%	-0.9%	8.4%
		p-value					0.14	0.33	0.14	0.90	0.96	0.63	0.40	0.81
F9000	AVL 8-mode	% change	3.2%	-6.7%	11.7%	-19.4%					-9.5%	-13.6%	8.6%	-36.6%
		p-value	0.72	0.01	0.16	0.05					0.21	0.02	0.22	0.04
Camp Pendleton Bus	AVL 8-mode	% change	1.3%	-6.8%	-0.4%	-10.8%	13.3%	-6.7%	-3.7%					
		p-value	0.96	0.29	0.91	0.28	0.62	0.20	0.29					
250 kW Generator	5-mode	% change	-6.6%	5.8%	2.3%	17.1%								
		p-value	0.20	0.25	0.21	0.48								
60 kW Generator	5-mode	% change	6.3%	13.0%	8.2%	10.9%								
		p-value	0.69	0.41	0.40	0.57								
Cheyenne Mountain Bus – UCR	Custom	% change					-22.0%	-17%	3.0%	-29%				
		p-value					0.18	0.01	0.12	0.00				
Cheyenne Mountain Bus- NREL	Custom	% change					-11.2%	-1.3%	0.2%	-9.4%				
		p-value					0.00	0.67	0.81	0.39				
Aircraft tow	In-use	% change	9%	-16%	-1%									
		p-value	0.91	0.31	0.83									
Forklift	In-use	% change					-10%	20%	-8%					
		p-value					**	**	**					
Model A1 Humvee	In-use	% change	NA	5%	6%									
		p-value	**	**	**									

Comparison of Engine and Vehicle Test Results

The results in EPA's 2002 compilation of published studies suggested a discrepancy between engine and vehicle tests for B20. In particular, engine tests as reviewed by EPA indicate a 2% increase in NO_x emissions for B20 while vehicle tests on average tend to indicate a smaller or even zero increase. However, a close examination of the data included in the review reveals that results are dominated by engines from a single manufacturer with a very limited range of model years. Nearly half of the observations (44%) were for DDC engines in the 1991 to 1997 model year range, and a large majority of these are for the Series 60 model. Engine manufacturers certified more than 700 heavy-duty engine families and 5,000 engine models in 2006 alone [34], although not all of these were on highway or diesel engines. In 2002 there were more than 5 million medium, light-heavy, and heavy-heavy duty trucks registered in the United States [35], and roughly 50% were 10 years old or older. These vehicles typically stay on the road for 15 years or longer. We believe that EPA's conclusion of a NO_x increase is influenced by the unrepresentative composition of the engine dataset. A hallmark of the B20 emission test results is that NO_x is highly variable, with percentage change ranging from roughly -7% to +7%. Data for the DDC Series 60 engine, which typically exhibits a small NO_x increase for B20, makes up a large fraction of the data reviewed. Therefore, EPA draws a conclusion that is at odds with the results of the more recent studies reviewed here.

An examination of all of the published data suggests that there is no discrepancy between engine and vehicle testing and that for B20 on average there is no net impact on NO_x. However, the reasons for the variability in NO_x with engine model are not understood and are worthy of further study. It is possible that the variation is caused by differences in how engine fuel injection systems and electronic controls respond to the lower energy content or other properties of B20.

Fundamental Studies of Biodiesel and NO_x Emissions

A combustion analysis study of biodiesel and biodiesel blends concluded that biodiesel blends had a shorter ignition delay than diesel alone, at both full and light load, and a lower premixed burn fraction at full load. However, diffusion burn rates were similar [36]. The shorter ignition delay, caused by biodiesel's higher cetane number, has been suggested as being the cause of the NO_x increase observed in many studies because the advanced combustion timing increases peak pressure and temperature. However, this is inconsistent with EPA's review of cetane number effects, which shows decreasing NO_x for increasing cetane number [37], although benefits are less for newer engines with more highly retarded injection timing. Use of cetane enhancing additives has been shown to reduce NO_x for B20 in older, more cetane-sensitive engines [15].

A number of other hypotheses on the cause of the increase in NO_x observed for biodiesel under some engine operating conditions have been advanced. Increasing NO_x may be caused by an increase in flame temperature in either premixed or diffusion burn, which is caused by reduction in the concentration of carbonaceous soot - a highly effective heat radiator. The net result of the PM reduction caused by supplying oxygen to the fuel rich zone of the diffusion flame may be to increase flame temperature because of this loss of radiant heat transfer [10]. This hypothesis has been investigated by Cheng and coworkers using an optically accessible engine and fuels with identical ignition delay [38]. NO_x emissions were higher for B100, even with matched ignition delay, especially at lower loads. Flame luminosity measurement suggested less radiation from the B100 flame, particularly under light load conditions where NO_x was shown to increase.

The double bonds present in biodiesel may cause a higher adiabatic flame temperature, and hence a higher temperature at the flame front in the diffusion flame. This hypothesis is consistent with results showing higher levels of NO_x emissions for biodiesel from more highly unsaturated feedstocks [14]. Cheng and coworkers present results of equilibrium calculations that refute this hypothesis [38]. However, Ban-Weiss and coworkers performed calculations of adiabatic flame temperature based on chemical kinetic models that suggest a significant impact of unsaturation [39].

A second fuel chemistry effect might be enhancement of the formation of prompt (or Fenimore) NO, which can account for up to 30% of NO_x formation [40]. Prompt NO is formed by reaction of radical HC species with nitrogen, ultimately leading to formation of NO. Hess and coworkers noted that unsaturated compounds may form higher levels of radicals during pyrolysis and combustion, and investigated the potential of radical scavenging antioxidant additives for NO_x reduction [41]. Some, but not all, antioxidants were shown to reduce NO_x emissions for their engine.

Van Gerpen and collaborators have shown that NO_x can increase as a result of a shift in fuel injection timing caused by different mechanical properties of biodiesel [42, 43]. Biodiesel has a higher bulk modulus of compressibility (or speed of sound) than petroleum diesel and this was proposed to cause a more rapid transfer of the fuel pump pressure wave to the injector needle. This caused earlier needle lift and a small advance in injection timing that was proposed to account for a fraction of the NO_x increase observed under some conditions. Sybist and Boehman also examined this effect [44]. They found that soy B100 produces a 1° advance in injection timing and a nearly 4° advance in the start of combustion. The bulk modulus effect appears to be applicable to pump-line nozzle and unit injection systems, but not for high-pressure common rail systems where "rapid transfer of a pressure wave" does not occur.

A number of more speculative hypotheses have been proposed. For example, reduction of the soot concentration in the flame may eliminate NO-carbon reactions. The importance of NO-carbon reactions in diesel combustion is unknown. Also, biodiesel has been shown to alter injection duration, spray properties, and other aspects of spray fluid flow [44]. The impact of these phenomena on NO_x emissions in this context is uncertain.

These studies indicate that there may be more than one factor contributing to the effect of biodiesel on NO_R. Furthermore, which factor is dominant may change with engine speed and load or with certain engine design parameters. Given the results of chassis and engine tests reviewed in previous sections, fundamental studies of biodiesel's impact on NO_R may be most relevant to B100, where it seems clear that an increase in NO_R occurs in most cases. Additional study is required to quantitatively understand the underlying factors causing biodiesel's impact on NO_R. Future studies should include a comparison of results from engine operating conditions where NO_R increases and where it does not.

METHODOLOGY

Vehicle Emissions Test Lab

All testing was conducted on a heavy-duty chassis dynamometer at NREL's ReFUEL test facility. The chassis dynamometer test facility includes analytical equipment for emissions and fuel economy measurements of on-road heavy-duty diesel vehicles. All emissions measurements are conducted in accordance with the Code of Federal Regulations (CFR), title 40, part 86, subpart N.

Chassis Dynamometer

The chassis dynamometer, as illustrated in Figure 5, is composed of three major components: the rolls – which are in direct contact with the vehicle tires during testing, the direct current (DC) electric motor (380 hp absorbing/360 hp motoring) dynamometer, and the flywheels. The DC electric motor and flywheels are installed in a pit below the ground level, such that the only exposed part of the dynamometer is the top of the 40– inch-diameter rolls. Two sets of rolls are installed, so that twin-axle vehicles can be tested. The dynamometer can simulate up to 80,000-lb vehicles at speeds up to 60 mph.

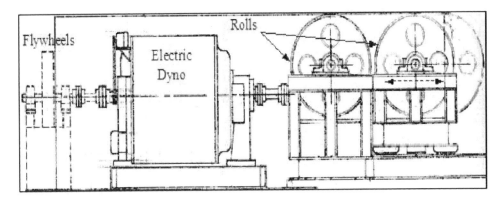

Figure 5. Chassis Dynamometer Schematic.

The rolls are the means by which power is absorbed from the vehicle. The rolls are attached to gearboxes that increase the speed of the central shaft by a factor of 5. The flywheels, mounted on the back of the dynamometer, provide a mechanical simulation of the vehicle inertia.

The energy absorption capability of the dynamometer is used to apply the road load, which is a summation of the aerodynamic drag and friction losses that the vehicle experiences, as a function of speed. The road load for each test vehicle was estimated from standard equations. The electric dynamometer is also used to adjust the simulated inertia, either higher or lower than the 31,000-lb base dynamometer inertia. The inertia simulation range of the chassis dynamometer is 8,000 to 80,000 lbs.

The test vehicle is secured with the drive axles over the rolls. A driver's aid monitor in the cab is used to guide the vehicle operator in driving the test trace. A large fan is used to cool the vehicle radiator during testing. The chassis dynamometer is supported by 72 channels of data acquisition, in addition to the emissions measurement, fuel metering, and combustion analysis subsystems.

With the vehicle jacked up off the rolls, an automated dynamometer warm-up procedure is performed daily, prior to testing, to ensure that parasitic losses in the dynamometer and gearboxes have stabilized at the appropriate level to provide repeatable loading. An unloaded coast-down procedure is also conducted to confirm that inertia and road load are being accurately simulated by the dynamometer control system. Between test runs a loaded coast-down procedure is performed to further ensure stability of vehicle and dynamometer parasitic losses and accurate road load simulation during testing.

Fuel Handling

Fuel supply from the vehicle's tank is interrupted, allowing for delivery of conditioned and metered fuel to the test vehicle. Test fuels are stored and blended, in drum quantities, in a temperature-conditioned shed. Test fuels are blended gravimetrically to the 20 volume percent level (B20). The fuel is delivered from the supply drum to the fuel metering and conditioning system, from which fuel is supplied to the vehicle's fueling system. The fuel metering system measures volumetric flow to an accuracy of +/- 0.5% of the reading. An in-line sensor measures the density with an accuracy of +/- 0.001 g/cc, allowing an accurate mass measurement over the test cycle.

Air Handling and Conditioning

Dilution air and the air supplied to the test vehicle for combustion are derived from a common source, a roof-mounted system that conditions the temperature of the air and humidifies as needed to meet desired specifications. This air is passed through a HEPA filter, in accordance with the (2007) CFR specifications, to eliminate background PM as a source of uncertainty in particulate measurements. The average inlet air temperature to the vehicle was maintained within a window of 24°C +/- 2°C for all test runs, and average humidity was controlled to 75 grains/lb (absolute) +/- 4 grains/lb.

Emissions Measurement

The emissions measurement system is based on the full-scale exhaust dilution tunnel method with a Constant Volume Sampling (CVS) system for mass flow measurement. The system is designed to comply with the requirements of the 2007 Code of Federal Regulations, title 40, part 86, subpart N. Exhaust from the vehicle flows through insulated piping to the full-scale 18-inch-diameter stainless steel dilution tunnel. A static mixer ensures thorough mixing of exhaust with conditioned, filtered, dilution air prior to sampling of the dilute exhaust stream to measure gaseous and particulate emissions.

A system with three Venturi nozzles is employed to maximize the flexibility of the emissions measurement system. Featuring 500-cfm, 1,000-cfm, and 1,500-cfm Venturi nozzles and gas-tight valves, the system flow can be varied from 500-cfm to 3,000-cfm flow rates in 500-cfm increments. This allows the dilution level to be tailored to the engine size being tested, maximizing the accuracy of the emissions measurement equipment.

The gaseous emissions bench is a Pierburg model AMA-2000. It features continuous analyzers for total HC, NO_x, CO, carbon dioxide (CO_2) and oxygen (O_2). The system features auto-ranging, automated calibration, zero check and span check features, as well as integration functions for calculating cycle emissions. There are two heated sample trains for gaseous emissions measurement: one for HC, and another for the other gaseous emissions. NO_x and HC measurements are performed on a wet basis, while CO, CO_2 and O_2, are done on a dry basis. Sample probes are located in the same plane in the dilution tunnel.

The PM sample control bench maintains a desired sample flow rate through the PM filters in proportion to the overall CVS flow, in accordance with the CFR. Stainless steel filter holders designed to the 2007 CFR requirements house 47-mm-diameter Teflon membrane filters through which the dilute exhaust sample flows.

Table 6. Description of Vehicles Tested

	Motor Coach	Freightliner Class 8	Conventional School Bus	Green Diesel School Bus	International Class 8	Transit Bus #1	Transit Bus #2	Transit Bus #3
Vehicle MY	Jan-04	May-99	Jul-04	Jan-06	Jan-06	Sep-00	Sep-00	Jun-00
Make	Sports Coach 37'	Freightliner	International	International	International	Orion	Orion	Orion
Odometer	33,320	503,468	30,441	2,274	3,165	136,610	205,387	108,451
Test Weight	23,500	64,000	26,000	26,000	64,000	35,000	35,000	35,000
Engine Manufacturer	Cummins	Detroit Diesel	International	International	Cummins	Cummins	Cummins	Cummins
Displacement	5.9 L	12.7 L	7.6 L	7.6 L	10.8 L	10.8 L	10.8 L	10.8 L
Engine MY	2003	2000	2004	2006	2005	2000	2000	2000
Engine Model	ISB 300	Series 60	D 285	DG 285	ISM 330	ISM 280	ISM 280	ISM 280
Rated HP	300	470	285	285	330	280	280	280
Test Fuels:	Certification, B20 Agland	Certification, B20	Certification, B20	Certification, B20	#2 Diesel, B20	#2 Diesel, B20	#2 Diesel, B20	#2 Diesel, B20
Petroleum	2007 Cert	2007 Cert	2007 Cert	2007 Cert	Local LSD C	Local LSD A	Local LSD A	Local LSD A & B
Biodiesel	Agland	Agland	Agland	Agland	Agland	BlueSun	BlueSun	Agland/BlueSun
Transmission Type	Allison Auto	Rockwell 10spd Manual	Allison Auto	Allison Auto	Eaton 10spd Manual	Friedrichshafen	Friedrichshafen	Friedrichshafen
Aftertreatment			DOC	DPF				

The PM sampling system is capable of drawing a sample directly from the large full-scale dilution tunnel or utilizing secondary dilution to achieve desired temperature, flow, and concentration characteristics. A cyclone separator, as described in the CFR requirements, is employed to mitigate tunnel PM artifacts. PM filters are handled, conditioned, and weighed in a Class 1000 clean room with precise control over the temperature and humidity (+/- 1°C for temperature and dew point). The microbalance for weighing PM filters features a readability of 0.1 μg (a CFR requirement), a barcode reader for filter identification and tracking, and a computer interface for data acquisition. The microbalance is installed on a specially designed table to eliminate variation in the measurement due to vibration.

Test Vehicles

This study includes data collected from chassis dynamometer testing of eight different heavy-duty on-road vehicles. Test vehicles included three transit buses, two school buses, two Class 8 trucks, and one motor coach. This collection of test vehicles captures a variety of engine makes, sizes, emissions control technologies and transmission types, but still cannot be considered as representative of on-road heavy-duty vehicles. Engine model year varied from 2000 to 2006. Accumulated mileage also varied for each of the test vehicles, ranging from 2,274 to 503,468 miles. Information detailing each of the test vehicles is provided in Table 6.

The three transit buses incorporated identical engine and transmission combinations, only differing in the accumulated mileage and biodiesel fuel type. This allowed for some assessment of the dependence of emission differences on vehicle-to-vehicle and fuel-to fuel variability.

Driving Cycles

Several different driving cycles were employed in this study. Driving cycles were chosen to mimic in-use operation for a given vehicle. The City-Suburban Heavy-Vehicle Cycle (CSHVC) was used for testing all but one of the vehicles in this study. This cycle, developed by West Virginia University (WVU), represents low-speed, stop-and-go driving events [45] and is shown in Figure 6. The Urban Dynamometer Driving Schedule for Heavy-Duty Vehicles (UDDS) was also employed. This cycle was developed from the same dataset used for development of the transient test portion of the heavy-duty FTP and is described in the Code of Federal Regulations [46], shown in Figure 7.

The Combined International Local and Commuter Cycle (CILCC) was developed by NREL for testing Class 4 to 6 hybrid electric delivery vehicles [47]. The only part of the cycle that is specific to hybrids is the length (>45 min). Otherwise, it is intended to simulate urban delivery driving for heavy-duty vehicles in general. The cycle was developed to use larger amounts of fuel energy so that changes in state of charge (battery energy) would be minimal in comparison. The acceleration events of this cycle were slightly modified to allow the Class 8 vehicle to achieve the drive trace; the cycle used is shown in Figure 8.

The Rowan University Composite School Bus Cycle (RUCSBC) [32], shown in Figure 9, was developed from school bus activity data. Note that the RUCSBC has the highest average and maximum acceleration rates and is therefore the most aggressive of these driving cycles. The Freeway Cycle [45] was developed from activity data on two heavy-duty trucks and includes four-lane highway driving with entrance and exit ramps, shown in Figure 10. The Freeway cycle has the highest average and maximum speed, and the longest distance. Cycle statistics for each of the drive cycles are shown in Table 7. The chosen drive cycles are representative of a wide range of driving styles, including high speed interstate driving to low-speed, stop-and-go driving.

Table 7. Cycle Statistics for Various Driving Cycles Used in this Study

	CSHVC	UDDS	CILCC mod	RUCSBC	Freeway
Total Time (sec)	1,700	1,060	3,192	1,310	1,640
Time at Idle (%)	23.24	36.32	15.57	21.15	9.27
Average Cycle Speed (mph)	14.15	18.81	14.25	20.95	34.03
Average Speed While Driving (mph)	18.44	29.56	16.89	26.59	37.52
Maximum Speed (mph)	43.8	58	55	49.7	60.7
Total Distance (mi)	6.68	5.54	12.64	7.63	15.51
Number of Stops (stops/mi)	1.95	1.26	1.98	1.44	0.58
Average Acceleration Rate (ft/sec^2)	1.31	1.57	1.44	2.1	0.67
Maximum Acceleration Rate (ft/sec^2)	3.81	6.01	3.67	12.17	4.69

Table 8. Properties of Test Fuels

	LSD A	B20 LSD A/	LSD B	B20 LSD B/	LSD C	B20 LSD C/	2007 Cert	B20 Cert/
Distillation T90, °C (D86)	325	340	310	331	316	333	299	327
Flash Point, °C (D93)	66	71	59	63	58	64	82	81
Copper Corrosion (D130)	1a	1a	1a	1a	1a	1a	--	1a
Kinematic Viscosity, cSt@40°C (D445)	2.438	2.726	2.247	2.548	2.382	2.687	2.3	2.527
Ash, %Mass (D482)	<0.001	<0.001	<0.001	<0.001	<0.001	<0.001	<0.001	<0.001
Carbon Residue, %mass (D524)	0.04	<0.010	0.11	0.08	0.13	0.11	--	0.09
Cetane Number (D613)	40.6	47	44.4	45.8	47.0	52.3	41	47.3
Cloud Point, °C (D5773)	-18	-14	-20	-28	-17	-13	--	-19
Total Sulfur, ppm (D5453)	364	280	320	264	304	245	12	8.6
Water & Sediment, %Vol (D2709)	0.010	0.010	0.01	0.01	0.010	0.010	--	0.010
Aromatics, %Vol (D1319)	--	28.5	25.0	--	23.8	--	28.8	--
Acid Number, mg KOH/gram (D664)	0.01	0.16	<0.05	<0.05	<0.05	<0.05	--	<0.05
Peroxide Number, ppm (D3703)	0	8.1	2.6	24.8	0	39.0	--	57.7

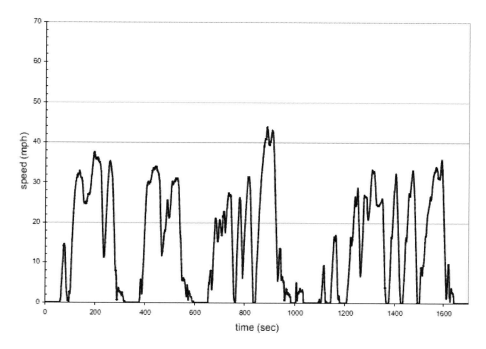

Figure 6. The CSHVC cycle.

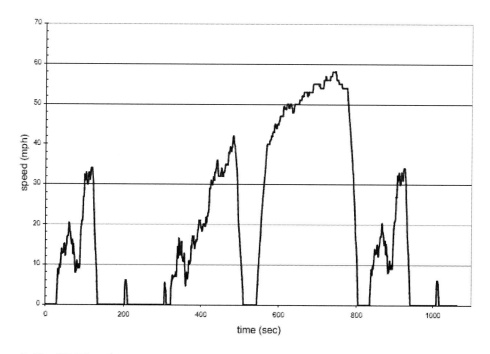

Figure 7. The UDDS cycle.

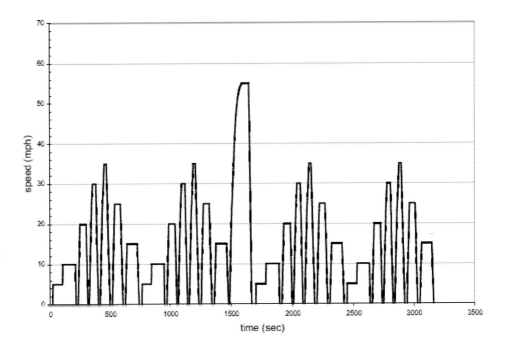

Figure 8. The CILCC modified cycle.

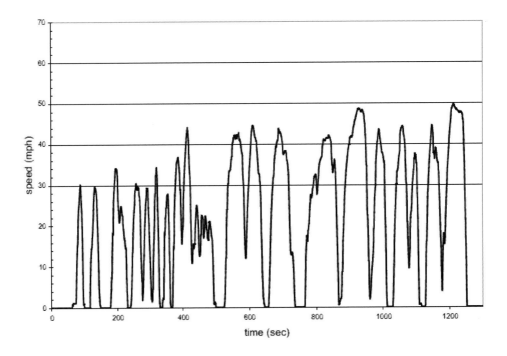

Figure 9. The RUCSBC cycle.

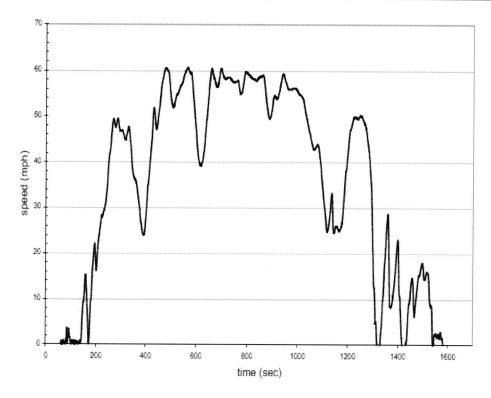

Figure 10. The Freeway cycle.

Test Fuels

Each vehicle was tested with a petroleum-derived diesel fuel and a 20% blend of soy-based biodiesel blended with the petroleum diesel base fuel (B20). In each case the B20 was splash blended on a volumetric basis. Two supplies of the soy-based biodiesel were used: a standard commercial grade fuel supplied by Agland, and a specialized biodiesel containing a proprietary multifunctional additive package supplied by BlueSun Biodiesel. Four separate supplies of petroleum diesel were used. Three on-highway low-sulfur diesels (LSD) were obtained locally, with LSD A and LSD B obtained from the local bus company at different times. LSD C was obtained from a local fuel jobber. The fourth fuel was ultra low sulfur 2007 certification diesel (2007 Cert) obtained from ChevronPhillips. Properties of these test fuels are listed in Table 8, while Table 6 notes which fuels were tested in each vehicle. The B20 blends were prepared using a highly accurate gravimetric procedure. Note that aromatic content is not reported for the B20 blends because the method used to measure fuel aromatic content, ASTM D1319, gives false high values for biodiesel and biodiesel blends. The aromatic content of B100 is zero, thus B20 blends will have 20% lower aromatic content than the diesel fuel in which they are blended.

RESULTS

Average results for each vehicle and drive cycle are shown in Tables 9 through 22. Percent differences as a result of biodiesel are also shown. Graphical representations of relative NO_x, PM, CO and HC emissions for each vehicle and drive cycle are shown in Figures 11 through 14. All results are averages of three or more individual runs, and a tabulation of individual run results is found in the Appendix.

Transit Bus Results

Data showing average emissions and fuel economy results for the three transit buses tested on the CSHVC are in Tables 9 through 12. Buses #1 and #2 were initially tested as part of a fleet evaluation project with the goal of measuring the effect of biodiesel usage with the actual in-use fuels [48]. Thus, these buses were tested on LSD A used by the bus company and the same fuel blended with soy biodiesel obtained from BlueSun Biodiesel and containing a proprietary multifunctional diesel additive. Both exhibit the roughly 2% fuel economy reduction expected for B20 based on fuel volumetric energy content. NO_x was reduced by 5.8% for bus #1 and by 3.9% for bus #2. To determine if the multifunctional additive was responsible for the NO_x reduction, a third, identical bus was tested using LSD A and biodiesel from BlueSun containing the additive LSD B with biodiesel from a second source (Agland) with no additive. These results are shown in Tables 11 and 12. Biodiesel from both sources produced a roughly 3 to 4% reduction in NO_x, suggesting that the NO_x reduction occurs generally for biodiesel for this engine-transmission combination on this drive cycle. However, because the base fuels are not identical, this is not a definitive comparison. All changes in NO_x are significant at 95% confidence or better. PM emission reductions were in the 15 to 20% range for all three buses with all changes significant at 90% confidence or better.

Table 9. Emission Test Results for Transit Bus #1 on CSHVC
Comparing LSD A and B20/BlueSun Biodiesel

	NOx (g/mile)	PM (g/mile)	CO (g/mile)	THC (g/mile)	Fuel Econ (mpg)	Fuel Cons (g/mile)
Diesel 95% conf	19.80 *0.34*	0.2740 0.0333	3.60 0.31	0.871 0.071	4.67 0.07	688 *11*
B20 95% conf	18.65 *0.15*	0.2264 0.0195	2.63 0.22	0.625 0.080	4.56 0.08	708 *9*
% Difference	-5.8%	-17.4%	-26.8%	-28.3%	-2.2%	2.9%
p-value	0.0001	0.0363	0.0006	0.0011	0.0809	0.0214

Table 10. Emission Test Results for Transit Bus #2 on CSHVC Comparing LSD A and B20/BlueSun Biodiesel

	NOx (g/mile)	PM (g/mile)	CO (g/mile)	THC (g/mile)	Fuel Econ (mpg)	Fuel Cons (g/mile)
Diesel	19.44	0.3210	3.43	0.794	4.54	709
95% conf	0.41	0.1170	0.47	0.065	0.13	21
B20	18.67	0.2150	2.73	0.571	4.45	730
95% conf	0.26	0.0393	0.32	0.022	0.09	15
% Difference	-3.9%	-33.0%	-20.3%	-28.0%	-2.0%	3.0%
p-value	0.0073	0.0832	0.0276	0.0001	0.2635	0.1304

Table 11. Emission Test Results for Transit Bus #3 on CSHVC Comparing LSD A and B20/BlueSun Biodiesel

	NOx (g/mile)	PM (g/mile)	CO (g/mile)	THC (g/mile)	Fuel Econ (mpg)	Fuel Cons (g/mile)
Diesel 95% conf	19.78 *0.17*	0.3079 0.0267	3.04 0.14	0.824 0.018	4.60 0.02	695 *3*
B20 95% conf	19.04 *0.15*	0.2447 0.0125	2.48 0.18	0.592 0.046	4.51 0.04	715 7
% Difference	-3.7%	-20.5%	-18.6%	-28.1%	-1.9%	2.8%
p-value	0.0001	0.0018	0.0007	0.0001	0.0044	0.0005

Table 12. Emission Test Results for Transit Bus #3 on CSHVC Comparing LSD B and B20/Agland Biodiesel

	NOx (g/mile)	PM (g/mile)	CO (g/mile)	THC (g/mile)	Fuel Econ (mpg)	Fuel Cons (g/mile)
Diesel	20.24	0.2805	3.07	0.824	4.59	696
95% conf	0.26	0.0252	0.26	0.017	0.04	6
B20	19.70	0.2324	2.70	0.659	4.50	716
95% conf	0.28	0.0100	0.17	0.049	0.04	6
% Difference	-2.7%	-17.2%	-11.9%	-20.0%	-1.9%	2.8%
p-value	0.0185	0.0109	0.0423	0.0001	0.0124	0.0014

Class 8 Truck Results

Results for the International Class 8 truck are shown in Tables 13 and 14. This vehicle was tested as a baseline vehicle for a heavy-duty hybrid electric vehicle study. Thus the CILCC, which was developed to simulate urban driving generally but with features designed to exercise hybrid vehicles, was used. Additionally, the Freeway cycle was used so that both city and freeway driving were simulated. This vehicle exhibited no significant change in NOx for the stop and go CILCC but a 2.3% increase in NO$_x$ for freeway driving (p<0.05). PM emission reductions on both cycles were quite high, on the order of 30%. Fuel economy reduction was the expected 2% on the CILCC but only 0.5% on the Freeway cycle.

Results for the Freightliner Class 8 truck are shown in Tables 15 and 16. This vehicle was tested exclusively for this study of biodiesel emissions and the CSHVC and Freeway cycles were employed. NO$_x$ emissions increased 2.1% and 3.6% on these cycles, respectively (p<0.05). PM emission reductions were in the 20 to 25% range. Fuel economy reduction was about 1.5% on both cycles.

Table 13. Emission Test Results for International Class 8 on CILCCmod Comparing LSD C and B20/Agland Biodiesel

	NOx (g/mile)	PM (g/mile)	CO (g/mile)	THC (g/mile)	Fuel Econ (mpg)	Fuel Cons (g/mile)
Diesel	11.04	0.2890	4.98	1.192	4.32	740
95% conf	0.14	0.0083	0.19	0.032	0.05	12
B20	11.03	0.2103	4.22	0.992	4.22	762
95% conf	0.19	0.0052	0.09	0.034	0.04	7
% Difference	-0.1%	-27.2%	-15.3%	-16.8%	-2.3%	2.9%
p-value	0.9528	0.0001	0.0020	0.0011	0.0402	0.0429

Table 14. Emission Test Results for International Class 8 on Freeway Cycle Comparing LSD C and B20/Agland Biodiesel

	NOx (g/mile)	PM (g/mile)	CO (g/mile)	THC (g/mile)	Fuel Econ (mpg)	Fuel Cons (g/mile)
Diesel 95% conf	6.75 0.02	0.2163 0.0104	2.13 0.04	0.515 0.003	5.44 0.02	586 3
B20 95% conf	6.90 0.10	0.1412 0.0010	1.82 0.03	0.452 0.009	5.41 0.03	594 3
% Difference	2.3%	-34.7%	-14.5%	-12.4%	-0.5%	1.4%
p-value	0.0340	0.0001	0.0002	0.0002	0.2410	0.0180

Table 15. Emission Test Results for the Freightliner Class 8 on CSHVC Comparing 2007 Certification Diesel and B20/Agland Biodiesel

	NOx (g/mile)	PM (g/mile)	CO (g/mile)	THC (g/mile)	Fuel Econ (mpg)	Fuel Cons (g/mile)
Diesel 95% conf	29.65 0.40	1.8303 0.2139	27.41 1.51	0.536 0.022	3.49 0.04	913 11
B20 95% conf	30.26 0.32	1.4761 0.0821	24.49 20.3	0.454 0.019	3.44 0.04	935 12
% Difference	2.1%	-19.4%	-10.7%	-15.2%	-1.5%	2.4%
p-value	0.0412	0.0129	0.0867	0.0003	0.1283	0.0253

Table 16. Emission Test Results for the Freightliner Class 8 on the Freeway Cycle Comparing 2007 Certification Diesel and B20/Agland Biodiesel

	NOx (g/mile)	PM (g/mile)	CO (g/mile)	THC (g/mile)	Fuel Econ (mpg)	Fuel Cons (g/mile)
Diesel	22.27	0.4826	8.14	0.200	5.90	539
95% conf	0.36	0.0650	0.29	0.013	0.03	3
B20	23.08	0.3563	7.58	0.168	5.81	553
95% conf	0.37	0.0219	0.12	0.014	0.03	2
% Difference	3.6%	-26.2%	-6.9%	-16.0%	-1.6%	2.6%
p-value	0.0124	0.0048	0.0058	0.0095	0.0007	0.0001

Motor Coach Results

The motor coach (or recreational vehicle) was tested on the CSHVC and UDDS cycles. This vehicle exhibited a roughly 3% increase in NO_x and 30% reduction in PM for both cycles. Fuel economy reduction was roughly 1%.

Table 17. Emission Test Results for the Motor Coach on CSHVC Comparing 2007 Certification Diesel and B20/Agland Biodiesel

	NOx (g/mile)	PM (g/mile)	CO (g/mile)	THC (g/mile)	Fuel Econ (mpg)	Fuel Cons (g/mile)
Diesel 95% conf	7.75 0.11	0.2538 0.0179	4.05 0.31	0.228 0.019	6.63 0.03	485 2
B20 95% conf	7.96 0.13	0.1825 0.0058	3.15 0.15	0.195 0.007	6.54 0.03	495 3
% Difference	2.8%	-28.1%	-22.3%	-14.5%	-1.3%	2.0%
p-value	0.0368	0.0001	0.0005	0.0092	0.0048	0.0002

Table 18. Emission Test Results for the Motor Coach on the UDDS Comparing 2007 Certification Diesel and B20/Agland Biodiesel

	NOx (g/mile)	PM (g/mile)	CO (g/mile)	THC (g/mile)	Fuel Econ (mpg)	Fuel Cons (g/mile)
Diesel	6.99	0.2387	3.66	0.138	7.05	456
95% conf	0.10	0.0079	0.18	0.014	0.18	12
B20	7.22	0.1672	2.95	0.133	7.00	462
95% conf	0.19	0.0128	0.09	0.019	0.09	6
% Difference	3.4%	-30.0%	-19.2%	-3.4%	-0.6%	1.4%
p-value	0.0576	0.0001	0.0001	0.6993	0.6734	0.3700

School Bus Results

The two school buses tested in this study were the only vehicles equipped with exhaust aftertreatment devices. The conventional International school bus was equipped with a diesel oxidation catalyst and the International Green Diesel school bus was equipped with a diesel particle filter (DPF). Results for the Green Diesel school bus are shown in Tables 19 and 20 for the CSHVC and the highly aggressive RUCSBC, respectively. NO$_x$ emissions were essentially unchanged on the CSHVC but increased by 2.3% for the RUCSBC. The DPF was highly effective at reducing PM emissions with values below 0.002 g/mile in all cases. This is roughly a factor of 100 below PM emissions measured for the other vehicles in this study. While examination of percent change in PM emissions for B20 suggests that PM has increased, the actual magnitude of these changes is extremely small and not statistically significant (p>0.05). Fuel economy was decreased by 1 to 2% for B20 in this vehicle.

Results for the conventional school bus are shown in Tables 21 and 22. This bus exhibited much more highly variable emissions than any of the other vehicles tested, reducing our ability to make definitive statements about emission differences. Examination of individual run results in the Appendix indicates some difficulty in controlling intake air humidity for this test sequence, but also shows large shifts in PM emissions with no apparent cause. Results are no change in NO$_x$ for the CSHVC but a 6.2% increase for the RUCSBC. PM emissions were unchanged for the CSHVC but decreased by 24% for the RUCSBC. Fuel economy declined by up to 1%.

Table 19. Emission Test Results for the International Green Diesel School Bus on CSHVC Comparing 2007 Certification Diesel and B20/Agland Biodiesel

	NOx (g/mile)	PM (g/mile)	CO (g/mile)	THC (g/mile)	Fuel Econ (mpg)	Fuel Cons (g/mile)
Diesel	7.70	0.0009	0.15	0.023	5.86	549
95% conf	0.14	0.0002	0.07	0.015	0.06	6
B20	7.64	0.0012	0.12	0.031	5.74	565
95% conf	0.09	0.0001	0.05	0.008	0.06	6
% Difference	-0.8%	28.0%	-15.9%	35.2%	-2.0%	2.8%
p-value	0.5484	0.1032	0.5158	0.7179	0.0328	0.0051

Table 20. Emission Test Results for the International Green Diesel School Bus n RUCSBC Comparing 2007 Certification Diesel and B20/Agland Biodiesel

	NOx (g/mile)	PM (g/mile)	CO (g/mile)	THC (g/mile)	Fuel Econ (mpg)	Fuel Cons (g/mile)
Diesel	8.93	0.0014	0.10	0.023	4.97	648
95% conf	0.08	0.0002	0.03	0.005	0.03	4
B20	9.14	0.0017	0.06	0.021	4.93	659
95% conf	0.16	0.0003	0.02	0.009	0.04	5
% Difference	2.3%	15.6%	-41.7%	-7.0%	-0.8%	1.7%
p-value	0.0346	0.2209	0.0547	0.7331	0.1561	0.0081

Table 21. Emission Test Results for the International Conventional School Bus on CSHVC Comparing 2007 Certification Diesel and B20/Agland Biodiesel

	NOx (g/mile)	PM (g/mile)	CO (g/mile)	THC (g/mile)	Fuel Econ (mpg)	Fuel Cons (g/mile)
Diesel 95% conf	9.85 0.10	0.1929 0.0210	5.22 0.37	0.439 0.041	5.93 0.04	534 3
B20 95% conf	9.79 0.12	0.1977 0.0176	5.72 0.46	0.434 0.047	5.86 0.03	549 3
% Difference	-0.7%	2.5%	9.5%	-1.1%	-1.1%	2.7%
p-value	0.4145	0.7368	0.1198	0.8802	0.0074	0.0001

Table 22. Emission Test Results for the International Conventional School Bus on RUCSBC Comparing 2007 Certification Diesel and B20/Agland Biodiesel

	NOx (g/mile)	PM (g/mile)	CO (g/mile)	THC (g/mile)	Fuel Econ (mpg)	Fuel Cons (g/mile)
Diesel	9.78	0.6954	8.95	0.373	5.01	633
95% conf	0.12	0.0324	0.49	0.074	0.03	4
B20	10.39	0.5284	6.93	0.300	4.99	645
95% conf	0.17	0.0393	1.12	0.100	0.05	6
% Difference	6.2%	-24.0%	-22.6%	-19.6%	-0.3%	1.9%
p-value	0.0001	0.0001	0.0014	0.2665	0.5937	0.0049

Results Summary

Figures 11 through 14 summarize the results for NO_R, PM, CO, and HC (or total hydrocarbon, THC), respectively. As can be seen in the data, not only is the impact of B20 on NO_R emissions highly dependent on the test vehicle, but it is also dependent on the chosen drive cycle. All three of the transit buses demonstrated reductions in NO_R emissions, regardless of biodiesel supply. The motor coach and the Freightliner Class 8 truck both showed increases in NO_R emissions over each of their test cycles. The International Class 8 and both school buses showed increases over one test cycle, and reductions or no change over the other test cycle. However, PM emission reductions are quite robust, independent of technology and driving cycle with the exception of the DPFequipped vehicle.

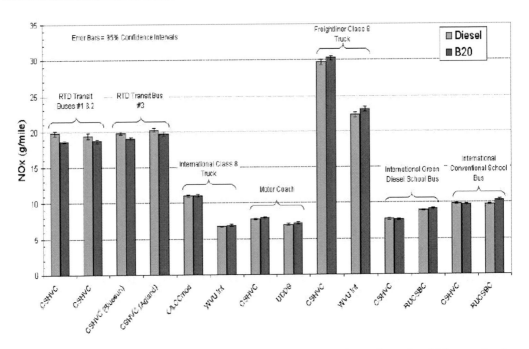

Figure 11. Comparison of NOx emissions for conventional diesel and B20 for each vehicle tested and each cycle.

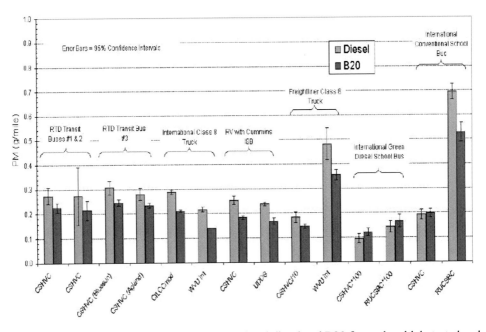

Figure 12. Comparison of PM emissions for conventional diesel and B20 for each vehicle tested and each cycle.

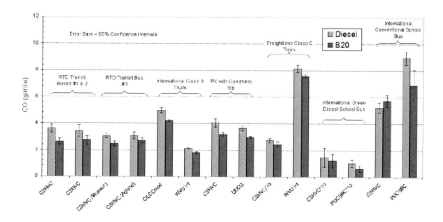

Figure 13. Comparison of CO emissions for conventional diesel and B20 for each vehicle tested and each cycle.

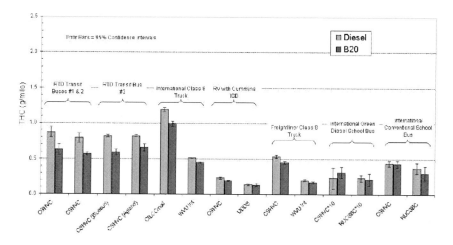

Figure 14. Comparison of THC emissions for conventional diesel and B20 for each vehicle tested and each cycle.

DISCUSSION

The average percent change in emissions and fuel economy for each vehicle and drive cycle are summarized in Table 23. Note that for Transit Bus #3 results for both B20A and B20B have been averaged so that this vehicle is not counted twice in the average. Across all vehicle/drive cycle combinations PM, CO and THC showed average reductions of 16.4%, 17.1%, and 11.6% respectively. NO_R increased on average by 0.6%. Fuel economy was reduced by an average of 1.4%. Table 23 also shows 95% confidence limits for these average values. Note that the confidence interval for NO_R emissions includes zero, or no change in NO_R. It is important to keep in mind, however, that this eight-vehicle dataset cannot in any way be considered as representative of in-use heavy-duty vehicles, or even of model year 2000 to 2006 vehicles. Nevertheless, the results confirm the robustness of PM, CO, and HC reductions found in most other studies, and support the conclusion that the impact of B20 on NO_R is not significant.

Additionally, if the results in Table 23 are combined with the soy B20 results from Tables 4 and 5, the average change in NO_R is 0.9%±1.5% (95% confidence interval).

Table 23. Average Percent Change in Emissions and Fuel Economy for All Vehicles Tested

Vehicle	Cycle	NOx % Change	PM % Change	CO % Change	THC% Change	Fuel Econ % Change
Transit Bus #1	CSHVC	-5.8	-17.4	-26.8	-28.3	-2.2
Transit Bus #2	CSHVC	-3.9	-33.0	-20.3	-28.0	-2.0
Transit Bus #3 (Average)	CSHVC	-3.2	-18.9	-15.3	-25.1	-1.9
Freightliner	CSHVC	2.1	-19.4	-10.7	-15.2	-1.5
Class 8	Freeway	3.6	-26.2	-6.9	-16.0	-1.6
Motor Coach	CSHVC	2.8	-28.1	-22.3	-14.5	-1.3
	UDDS	3.4	-30.0	-19.2	-3.4	-0.6
International	CILCCmod	-0.1	-27.2	-15.3	-16.8	-2.3
Class 8	Freeway	2.3	-34.7	11.5	12.4	0.5
Green Diesel	CSHVC	-0.8	28.0	-15.9	35.2	-2.0
School Bus	RUCSBC	2.3	15.6	-41.7	-7.0	-0.8
Conventional	CSHVC	-0.7	2.5	9.5	-1.1	-1.1
School Bus	RUCSBC	6.2	-24.0	-22.6	-19.6	-0.3
Overall Average % Difference		0.6	-16.4	-17.1	-11.6	-1.4
95% Confidence Interval		±1.8	±10	±6.1	±8.6	±0.36

Table 24 shows average change in emissions and fuel economy for subsets of the overall dataset. The vehicles tested include four meeting the 4-g/bhp-h NO_x requirement that went into effect in 1998, and four meeting the 2.5-g/bhp-h NO_x+HC requirement that went into effect in 2004 (or as early as 2002 for some manufacturers). The first two rows of Table 24 examine average emission changes for B20 in vehicles from these two technology groups. The most obvious observation is the reduction in NO_x observed for the 4-g/bhp-h engines compared to the increase observed for 2.5-g/bhp-h engines. However, three out of the four 4-g/bhp-h NO_x vehicles were identical transit buses. Comparisons made with this subset of vehicles may therefore not be applicable to 4-g/bhp-h NO_x vehicles in general and again highlight the fact that this small group of vehicles is not a representative sample.

Table 24. Average Percent Change in Emissions for Specific Subsets of the Total Dataset, 95% Confidence Interval is Shown

Vehicle Cycle	NOx % Change	PM % Change	CO % Change	THC% Change	Fuel Econ % Change
4.0-g/bhp-h Engines Only	-1.4±3.3	-23.0±5.2	-16.0±6.3	-22.3±5.1	-1.8±0.2
2.5-g/bhp-h Engines Only	1.9±1.7	-12.2±17	-17.8±9.7	-5.0±12	-1.1±0.5
CSHVC Only	-1.4±2.2	-12.3±15	-14.5±8.2	-10.9±16	-1.7±0.3

All but one of the vehicles was driven on the CSHVC, thus it is of interest to examine results for this urban/suburban driving cycle separately, and the results are shown in Table 24. Percent changes for the CSHVC are quite similar to the changes observed overall. However, for the CSHVC, NO_x emissions decrease slightly. A number of other comparisons might be made, for example comparing emission changes for LSD versus ULSD, but the eight-vehicle dataset presented here is too small for meaningful comparisons of this type to be made.

Examination of Real Time NO_x Emissions Data

The impact of biodiesel on NO_x emissions varies with vehicle, engine technology, and chosen drive cycle. An analysis of real-time NO_x data illustrates this impact relative to different driving events. Figures 15 through 19 show snapshots of real-time NO_x data for portions of various drive traces and with different vehicles. In each case, the data is presented for both test fuels in order to show comparisons of how biodiesel impacts NO_x emissions through different driving events.

Figure 15 shows NO_x traces for a portion of the CSHVC cycle driven by RTD transit bus #3. This is a 4.0-g/bhp-h NO_x engine, thus it does not incorporate EGR for NO_x control. NO_x emissions, shown in grams/second, differ under several driving events. During idle portions of the drive cycle, B20 causes a significant decrease in NO_x emissions. During most acceleration events, the peaks in NO_x emissions are higher for B20, particularly at or just before peak speed. However, for some acceleration events NO_x is lower for B20, especially during longer accelerations but before peak speed (i.e., 1,140 to 1,150 seconds and 1,210 to 1,220 seconds in Figure 15). The combination of these effects causes overall NO_x emissions to decrease with the use of B20 in this vehicle. The 4-g/bhp-h Freightliner Class 8 truck (Figure 16) shows no difference at idle or during acceleration, but higher NO_x for B20 at speed peaks.

Figure 17 shows real-time NO_x traces for the same portion of the CSHVC cycle, driven by the motor coach. This is a 2.5-g/bhp-h NO_x engine, thus incorporating EGR. As can be seen, NO_x emissions at the idle conditions are controlled to nearly zero grams/second, thus there is no difference in NO_x emissions between the two fuels. However, this vehicle still experiences the higher peaks in NO_x emissions with B20 under acceleration events or near peak speed, leading to an overall increase in NO_x emissions.

Figure 18 shows the real-time NO_x traces for the International Class 8 Truck driven over the CILCC drive cycle. Like the motor coach, this vehicle also incorporates a 2.5 g/bhp-h NO_x engine. However, this engine shows slight decreases in the NO_x peaks during the acceleration events. NO_x emissions during steady-state and idle operation are the same. The overall NO_x emissions for the International Class 8 truck on this drive cycle showed slight reductions with B20, but not with statistical confidence ($p = 0.9528$).

These results do not reveal any obvious, consistent factor that is causing the variability observed for NO_x with these vehicles. A much more detailed analysis will be required. In particular we recommend an analysis that examines factors such as acceleration rate, wheel horsepower, and rate of change of horsepower. A study that employs a transmission model to estimate engine torque at various driving conditions may also prove valuable [49].

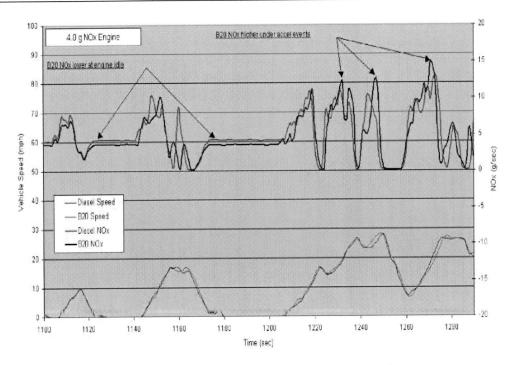

Figure 15. Portions of the CSHVC real-time NO$_x$ traces for RTD Transit Bus #3.

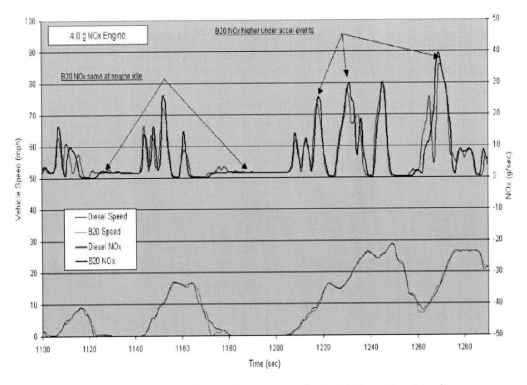

Figure 16. Portions of the CSHVC real-time NO$_x$ traces for the Freightliner Class 8 truck.

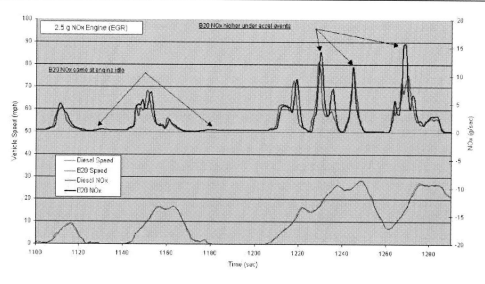

Figure 17. Portions of the CSHVC real-time NO$_x$ traces for the motor coach.

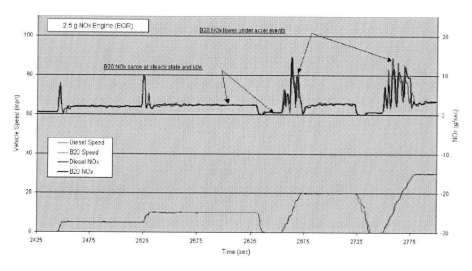

Figure 18. Portions of the CILCC real-time NO$_x$ traces for the International Class 8 truck.

SUMMARY AND RECOMMENDATIONS

The objective of this study was to determine if testing entire vehicles on a heavy-duty chassis dynamometer provides a better, more realistic measurement of the impact of B20 on regulated pollutant emissions. This report also documents completion of the National Renewable Energy Laboratory's Fiscal Year 2006 Annual Operating Plan Milestone 10.4. This milestone supports the U.S. Department of Energy, Fuels Technologies Program Multiyear Program Plan Goal of identifying fuels that can displace 5% of petroleum diesel by 2010.

An EPA review of engine testing studies on biodiesel concluded that on average, for soy biodiesel, NO$_x$ emissions increase by 2% [16]. Careful examination of the test data on which this conclusion is based shows that nearly half of the observations (44%) were for DDC engines in the 1991 to 1997 model year range, and a large majority of these are for the Series 60 model. We believe that EPA's conclusion of a NO_x increase is influenced by the unrepresentative composition of the engine dataset. A hallmark of the B20 emission test results is that NO_x is highly variable, with percentage change ranging from roughly -7% to +7%. Because data for the DDC Series 60 engine, which typically exhibits a small NO_x increase for B20, makes up such a large fraction of the data reviewed, EPA draws a conclusion that is at odds with the results of more recent studies.

Here we review more recently published studies (Table 3) and find an average change in NO_x for all recent B20 studies of -0.6%±2.0% (95% confidence interval). Restricting the average to recent studies of B20 with soy biodiesel yields an average NO_x impact of 0.1%±2.7%. The EPA review also includes summary of a smaller vehicle testing dataset that shows no significant impact of biodiesel on NO_x. We reviewed several more recently published vehicle (chassis) testing studies (Tables 4 and 5) and found an average change in NO_x of 1.2%±2.9% (95% confidence interval).

In the work reported here, eight heavy-duty diesel vehicles were tested, including three transit buses, two school buses, two Class 8 trucks, and one motor coach. Four of these vehicles met the 1998 heavy-duty emissions requirement of 4 g/bhp-h NO_x and four met the 2004 limit of 2.5 g/bhp-h NO_x+HC. Driving cycles that simulate both urban and freeway driving were employed. Each vehicle was tested on a petroleum-derived diesel fuel and on a 20 volume percent blend of that fuel with soy derived biodiesel. On average B20 caused PM and CO emissions to be reduced by 16% to 17% and HC emissions to be reduced by 12% relative to petroleum diesel. Emissions of these three regulated pollutants nearly always went down, the one exception being a vehicle equipped with a DPF that showed very low emissions of PM, CO, and HC. Furthermore, there was no significant change in these emissions for blending of B20. The NO_x emissions impact of B20 varied widely with engine/vehicle technology and test cycle ranging from -5.8% to +6.2%. On average, NO_x emissions did not change (statistically insignificant 0.6% average change). If the results of this study are combined with the soy B20 results from Tables 4 and 5 (recently published studies), the average change in NO_x is 0.9%±1.5% (95% confidence interval).

Based on the studies reviewed and the new data reported here, there does not appear to be a discrepancy between engine and chassis testing studies for the effect of B20 on NO_x emissions. The apparent disagreement that exists between engine testing results and chassis testing results in EPA's 2002 review occurred because neither of these datasets is representative of the on-road fleet. Newer studies are not more representative, but if all of the available data are viewed together we conclude that B20 has no significant impact on NO_x.

A preliminary examination of real-time NO_x emissions data did not reveal any consistent reason for the wide range in NO_x emission results for different vehicles. It is recommended that the real-time data be more fully analyzed in a study that considers the effect of vehicle speed and acceleration, as well as wheel horsepower and rate of change of horsepower. Additionally, modeling of the vehicle transmission to estimate actual engine torque output is recommended. Given the significant amount of additional data now available, an updating and revision of the EPA review is also recommended. And it is further recommended that strict quality criteria be applied, and studies with inadequate documentation, methodology, or controls be rejected.

APPENDIX: DETAILED CHASSIS TEST DATA

Table 25. RTD Transit Bus #1 – CSHVC – LSD A

Date		Run	Driver	CO2 (g/mile)	NOx (g/mile)	THC (g/mile)	CO (g/mile)	PM (g/mile)	Distance (miles)	Fuel Cons (g/mile)	Fuel Econ (mpg)	Humidity (grain/lb)	Temp (F)
2/1/2005	Base	332	John	2002	19.90	0.906	4.18	0.2380	6.65	686	4.68	72	69
2/1/2005	Base	333	Stuart	2007	20.56	1.033	3.74	0.2185	6.63	683	4.70	74	68
2/2/2005	Base	337	John	2058	19.69	0.816	3.48	0.3055	6.65	714	4.50	67	73
2/2/2005	Base	338	Stuart	2023	19.81	0.854	3.08	0.2616	6.64	686	4.68	69	74
2/2/2005	Base	339	John	2031	19.33	0.789	3.78	0.3263	6.68	688	4.67	69	74
2/2/2005	Base	340	Stuart	1971	19.54	0.829	3.32	0.2940	6.67	672	4.77	71	75
Average				2015	19.80	0.871	3.60	0.2740	6.65	688	4.67	70	72
Standard Deviation				29	0.42	0.089	0.39	0.0416	0.02	14	0.09	3	3
Coefficient of Variation				1.5%	2.1%	10.2%	10.8%	15.2%	0.3%	2.0%	1.9%	3.6%	4.3%

Table 26. RTD Transit Bus #1 – CSHVC – B20 (LSD A and BlueSun Biodiesel)

Date	Fuel	Run	Driver	CO2 (g/mile)	NOx (g/mile)	THC (g/mile)	CO (g/mile)	PM (g/mile)	Distance (miles)	Fuel Cons (g/mile)	Fuel Econ (mpg)	Humidity (grain/lb)	Temp (F)
2/1/2005	B20	326	Stuart	2,003	18.78	0.604	2.48	0.2421	6.62	719	4.43	69	73
2/1/2005	B20	328	John	1,979	18.54	0.598	2.94	0.2200	6.66	712	4.54	70	73
2/1/2005	B20	329	John	2,013	18.83	0.581	2.69	0.2365	6.67	719	4.50	69	74
2/2/2005	B20	345	John	2,012	18.36	0.485	2.97	0.2603	6.71	712	4.55	69	76
2/2/2005	B20	346	Stuart	1,985	18.83	0.729	2.35	0.1961	6.64	696	4.65	70	76
2/2/2005	B20	347	Stuart	1,952	18.57	0.754	2.36	0.2037	6.68	691	4.69	71	76
Average				1,991	18.65	0.625	2.63	0.2264	6.66	708	4.56	70	75
Standard Deviation				24	0.19	0.100	0.28	0.0244	0.03	12	0.09	1	2
Coefficient of Variation				1.2%	1.0%	16.0%	10.6%	10.8%	0.5%	1.7%	2.1%	0.9%	2.1%

Table 27. RTD Transit Bus #2 – CSHVC – LSD A

Date	Fuel	Run	Driver	CO2 (g/mile)	NOx (g/mile)	THC (g/mile)	CO (g/mile)	PM (g/mile)	Distance (miles)	Fuel Cons (g/mile)	Fuel Econ (mpg)	Humidity (grain/lb)	Temp (F)
2/4/2005	Base	364	Stuart	1,955	19.81	0.806	2.84	0.2178	6.65	689	4.67	71	73
2/4/2005	Base	365	Stuart	1,941	19.86	0.818	2.77	0.2001	6.68	689	4.67	71	73
2/4/2005	Base	366	John	1,975	19.24	0.724	3.64	0.2901	6.66	704	4.57	72	72
2/4/2005	Base	367	John	1,977	19.08	0.938	3.60	0.2804	6.66	699	4.60	73	71
2/7/2005	Base	371	Stuart	2,035	19.97	0.731	3.37	0.6017	6.64	760	4.23	69	73
2/7/2005	Base	372	John	1,968	18.68	0.747	4.35	0.3358	6.69	712	4.52	73	72
Average				1,975	19.44	0.794	3.43	0.3210	6.66	709	4.54	72	72
Standard Deviation				32	0.52	0.081	0.58	0.1462	0.02	26	0.16	2	1
Coefficient of Variation				1.6%	2.7%	10.2%	17.0%	45.6%	0.3%	3.7%	3.6%	2.3%	1.1%

Table 28. RTD Transit Bus #2 – CSHVC – B20 (LSD A and BlueSun Biodiesel)

Date	Fuel	Run	Driver	CO2 (g/mile)	NOx (g/mile)	THC (g/mile)	CO (g/mile)	PM (g/mile)	Distance (miles)	Fuel Cons (g/mile)	Fuel Econ (mpg)	Humidity (grain/lb)	Temp (F)
2/4/2005	B20	358	John	2,070	19.01	0.542	3.18	0.2583	6.66	765	4.24	73	73
2/4/2005	B20	359	Stuart	1,960	18.75	0.566	2.37	0.1933	6.65	750	4.33	71	75
2/4/2005	B20	360	John	1,990	18.47	0.546	2.93	0.2439	6.71	743	4.37	71	75
2/4/2005	B20	362	Stuart	1,967	19.17	0.564	2.43	0.1581	6.72	730	4.44	72	73
2/7/2005	B20	377	John	2,044	18.33	0.533	3.61	0.3261	6.70	729	4.45	74	75
2/7/2005	B20	378	Stuart	1,965	18.38	0.597	2.38	0.1868	6.68	707	4.59	74	73
2/7/2005	B20	379	John	1,954	18.20	0.616	2.53	0.1821	6.68	703	4.62	75	71
2/7/2005	B20	380	Stuart	1,978	19.08	0.608	2.42	0.1714	6.68	713	4.55	74	70
Average				1,991	18.67	0.571	2.73	0.2150	6.68	730	4.45	73	73
Standard Deviation				43	0.38	0.032	0.46	0.0568	0.02	22	0.13	2	2
Coefficient of Variation				2.1%	2.0%	5.6%	16.9%	26.4%	0.4%	3.0%	3.0%	2.2%	2.3%

Table 29. RTD Transit Bus #3 – CSHVC – LSD A

Date	Fuel	Run	Driver	CO2 (g/mile)	NOx (g/mile)	THC (g/mile)	CO (g/mile)	PM (g/mile)	Distance (miles)	Fuel Cons (g/mile)	Fuel Econ (mpg)	Humidity (grain/lb)	Temp (F)
9/8/2005	Base	725	Stuart	2,052	19.59	0.781	3.23	0.3421	6.69	693	4.61	73	75
9/8/2005	Base	726	Stuart	2,066	20.02	0.845	3.21	0.3457	6.68	697	4.59	78	77
9/8/2005	Base	727	Stuart	2,051	20.06	0.835	3.05	0.3144	6.68	689	4.64	68	80
9/12/2005	Base	752	Stuart	2,061	19.63	0.831	3.09	0.2954	6.68	700	4.57	76	75
9/12/2005	Base	753	Stuart	2,041	19.78	0.815	2.80	0.2579	6.67	699	4.57	72	76
9/12/2005	Base	754	Stuart	2,045	19.61	0.835	2.87	0.2919	6.67	696	4.59	76	76
Average				2,053	19.78	0.824	3.04	0.3079	6.68	695	4.60	73	78
Standard Deviation				9	0.21	0.023	0.18	0.0333	0.01	4	0.03	5	2
Coefficient of Variation				0.5%	1.1%	2.8%	5.8%	10.8%	0.1%	0.6%	0.6%	6.5%	3.1%

Table 30. RTD Transit Bus #3 – CSHVC – B20 (LSD A and BlueSun Biodiesel)

Date	Fuel	Run	Driver	CO2 (g/mile)	NOx (g/mile)	THC (g/mile)	CO (g/mile)	PM (g/mile)	Distance (miles)	Fuel Cons (g/mile)	Fuel Econ (mpg)	Humidity (grain/lb)	Temp (F)
9/8/2005	B20	729	Stuart	2,040	18.92	0.578	2.66	0.2511	6.69	708	4.55	75	76
9/8/2005	B20	730	Stuart	2,032	18.96	0.654	2.63	0.2615	6.67	708	4.55	76	79
9/8/2005	B20	731	Stuart	2,029	18.79	0.666	2.74	0.2612	6.67	706	4.56	78	77
9/12/2005	B20	749	Stuart	2,090	19.32	0.523	2.27	0.2379	6.67	723	4.47	73	74
9/12/2005	B20	750	Stuart	2,095	19.17	0.587	2.23	0.2235	6.69	724	4.46	73	74
9/12/2005	B20	751	Stuart	2,101	19.09	0.546	2.33	0.2332	6.69	723	4.47	74	75
Average				2,064	19.04	0.592	2.48	0.2447	6.68	715	4.51	76	77
Standard Deviation				34	0.19	0.057	0.22	0.0157	0.01	9	0.05	2	1
Coefficient of Variation				1.7%	1.0%	9.7%	9.0%	6.4%	0.2%	1.2%	1.1%	2.1%	1.6%

Table 31. RTD Transit Bus #3 – CSHVC – LSD B

Date	Fuel	Run	Driver	CO2 (g/mile)	NOx (g/mile)	THC (g/mile)	CO (g/mile)	PM (g/mile)	Distance (miles)	Fuel Cons (g/mile)	Fuel Econ (mpg)	Humidity (grain/lb)	Temp (F)
9/9/2005	Base	735	Stuart	2,055	20.40	0.817	2.69	0.2522	6.68	693	4.61	73	76
9/9/2005	Base	736	Stuart	2,042	20.27	0.834	2.80	0.2437	6.67	683	4.68	77	78
9/9/2005	Base	737	Stuart	2,068	20.74	0.818	2.87	0.2631	6.68	696	4.59	75	78
9/12/2005	Base	746	Stuart	2,086	20.20	0.860	3.36	0.3192	6.68	708	4.52	75	76
9/12/2005	Base	747	Stuart	2,032	19.78	0.803	3.45	0.3061	6.67	697	4.59	76	77
9/12/2005	Base	748	Stuart	2,057	20.05	0.810	3.25	0.2989	6.68	699	4.57	77	77
Average				2,057	20.24	0.824	3.07	0.2805	6.68	696	4.59	75	77
Standard Deviation				19	0.32	0.021	0.32	0.0314	0.01	8	0.05	2	1
Coefficient of Variation				0.9%	1.6%	2.5%	10.4%	11.2%	0.1%	1.2%	1.2%	2.6%	1.3%

Table 32. RTD Transit Bus #3 – CSHVC – B20 (LSD B and Agland Biodiesel)

Date	Fuel	Run	Driver	CO2 (g/mile)	NOx (g/mile)	THC (g/mile)	CO (g/mile)	PM (g/mile)	Distance (miles)	Fuel Cons (g/mile)	Fuel Econ (mpg)	Humidity (grain/lb)	Temp (F)
9/9/2005	B20	739	Stuart	2,032	19.94	0.685	2.69	0.2316	6.67	706	4.56	78	79
9/9/2005	B20	740	Stuart	2,029	20.07	0.719	2.50	0.2159	6.66	706	4.56	82	76
9/9/2005	B20	741	Stuart	2,040	19.96	0.695	2.46	nm	6.70	721	4.46	74	76
9/12/2005	B20	743	Stuart	2,087	19.48	0.545	2.91	0.2435	6.67	718	4.49	73	74
9/12/2005	B20	744	Stuart	2,055	19.54	0.667	2.66	0.2250	6.66	717	4.49	72	75
9/12/2005	B20	745	Stuart	2,041	19.19	0.644	2.99	0.2458	6.68	724	4.45	76	75
Average				2,047	19.70	0.659	2.70	0.2324	6.67	716	4.50	78	77
Standard Deviation				21	0.35	0.062	0.22	0.0126	0.02	8	0.05	4	2
Coefficient of Variation				1.0%	1.8%	9.3%	8.0%	5.4%	0.2%	1.1%	1.1%	5.2%	2.4%

Table 33. International Class 8 Truck – CILCCmod – LSD C

Date	Fuel	Run	Driver	CO2 (g/mile)	NOx (g/mile)	THC (g/mile)	CO (g/mile)	PM (g/mile)	Distance (miles)	Fuel Cons (g/mile)	Fuel Econ (mpg)	Humidity (grain/lb)	Temp (F)
04/19/06	Base	1,105	Greg	2,347	11.18	1.165	4.79	0.2844	12.70	749	4.28	75	74
04/19/06	Base	1,106	Greg	2,357	10.98	1.222	5.05	0.2975	12.67	744	4.31	78	74
04/19/06	Base	1,107	Greg	2,328	10.96	1.189	5.09	0.2851	12.72	728	4.37	74	76
Average				2,344	11.04	1.192	4.98	0.2890	12.70	740	4.32	75	75
Standard Deviation				15	0.12	0.029	0.16	0.0074	0.03	11	0.05	2	1
Coefficient of Variation				0.6%	1.1%	2.4%	3.3%	2.6%	0.2%	1.5%	1.0%	2.7%	1.4%

Table 34. International Class 8 Truck – CILCCmod – B20 (LSD C and Agland Biodiesel)

Date	Fuel	Run	Driver	CO2 (g/mile)	NOx (g/mile)	THC (g/mile)	CO (g/mile)	PM (g/mile)	Distance (miles)	Fuel Cons (g/mile)	Fuel Econ (mpg)	Humidity (grain/lb)	Temp (F)
04/19/06	B20	1,102	Greg	2,321	11.02	1.01	4.28	0.2141	12.72	760	4.23	72	73
04/19/06	B20	1,103	Greg	2,326	11.20	1.01	4.24	0.2052	12.73	756	4.25	72	74
04/19/06	B20	1,104	Greg	2,311	10.87	0.96	4.12	0.2116	12.75	769	4.18	73	74
Average				2,320	11.03	0.992	4.22	0.2103	12.73	762	4.22	72	73
Standard Deviation				8	0.16	0.030	0.08	0.0046	0.02	6	0.04	0	0
Coefficient of Variation				0.3%	1.5%	3.0%	2.0%	2.2%	0.1%	0.8%	0.9%	0.6%	0.6%

Table 35. International Class 8 Truck – Freeway Cycle – LSD C

Date	Fuel	Run	Driver	CO2 (g/mile)	NOx (g/mile)	THC (g/mile)	CO (g/mile)	PM (g/mile)	Distance (miles)	Fuel Cons (g/mile)	Fuel Econ (mpg)	Humidity (grain/lb)	Temp (F)
04/18/06	Base	1,093	Greg	1,899	6.73	0.51	2.13	0.2259	15.40	588	5.42	81	73
04/18/06	Base	1,094	Greg	1,897	6.74	0.52	2.10	0.2077	15.42	583	5.46	80	75
04/18/06	Base	1,095	Greg	1,904	6.76	0.51	2.16	0.2152	15.40	586	5.43	77	75
Average				1,900	6.75	0.515	2.13	0.2163	15.41	586	5.44	79	74
Standard Deviation				4	0.02	0.003	0.03	0.0092	0.01	2	0.02	2	1
Coefficient of Variation				0.2%	0.2%	0.5%	1.5%	4.2%	0.1%	0.4%	0.4%	2.5%	1.8%

Table 36. International Class 8 Truck – Freeway Cycle – B20 (LSD C and Agland Biodiesel)

Date	Fuel	Run	Driver	CO2 (g/mile)	NOx (g/mile)	THC (g/mile)	CO (g/mile)	PM (g/mile)	Distance (miles)	Fuel Cons (g/mile)	Fuel Econ (mpg)	Humidity (grain/lb)	Temp (F)
04/18/06	B20	1,097	Greg	1,884	7.00	0.45	1.80	0.1410	15.40	594	5.41	76	74
04/18/0	B20	1,099	Greg	1,867	6.85	0.44	1.85	0.1404	15.42	591	5.44	82	73
04/18/06	B20	1,100	Greg	1,883	6.86	0.46	1.80	0.1422	15.43	596	5.39	81	74
Average				1,878	6.90	0.452	1.82	0.1412	15.42	594	5.41	79	74
Standard Deviation				10	0.09	0.008	0.03	0.0009	0.02	3	0.02	3	0
Coefficient of Variation				0.5%	1.2%	1.8%	1.6%	0.6%	0.1%	0.4%	0.4%	4.1%	0.6%

Table 37. Freightliner Class 8 Truck – CSHVC – 2007 Cert Diesel

Date	Fuel	Run	Driver	CO2 (g/mile)	NOx (g/mile)	THC (g/mile)	CO (g/mile)	PM (g/mile)	Distance (miles)	Fuel Cons (g/mile)	Fuel Econ (mpg)	Humidity (grain/lb)	Temp (F)
06/28/06	Base	1,157	Greg	2,886	28.89	0.541	nm	2.1352	6.64	911	3.50	72	75
06/28/06	Base	1,158	Greg	2,927	29.64	0.517	nm	nm	6.64	923	3.45	70	75
06/28/06	Base	1,159	Greg	2,909	29.26	0.585	nm	nm	6.68	915	3.48	72	76
07/05/06	Base	1,167	Greg	2,910	29.79	0.543	28.83	1.8259	6.66	931	3.42	90	78
07/05/06	Base	1,168	Greg	2,796	30.16	0.519	26.18	1.6310	6.70	890	3.57	89	78
07/05/06	Base	1,169	Greg	2,848	30.15	0.509	27.21	1.7292	6.68	908	3.50	93	79
Average				2,879	29.65	0.536	27.41	1.8303	6.67	913	3.49	81	77
Standard Deviation				49	0.50	0.028	1.34	0.2183	0.02	14	0.05	11	2
Coefficient of Variation				1.7%	1.7%	5.2%	4.9%	11.9%	0.4%	1.5%	1.5%	13.3%	2.2%

Table 38. Freightliner Class 8 Truck – CSHVC – B20 (2007 Cert Diesel and Agland Biodiesel)

Date	Fuel	Run	Driver	CO2 (g/mile)	NOx (g/mile)	THC (g/mile)	CO (g/mile)	PM (g/mile)	Distance (miles)	Fuel Cons (g/mile)	Fuel Econ (mpg)	Humidity (grain/lb)	Temp (F)
06/28/06	B20	1,160	Greg	2,944	30.44	0.432	nm	nm	6.67	947	3.40	75	75
06/28/06	B20	1,161	Greg	2,860	29.94	0.438	nm	1.5571	6.67	924	3.48	73	77
06/28/06	B20	1,162	Greg	2,910	30.64	0.442	nm	1.4961	6.65	947	3.39	69	76
07/05/06	B20	1,164	Greg	2,951	30.73	0.493	23.29	1.4679	6.43	946	3.40	76	79
07/05/06	B20	1,165	Greg	2,827	30.02	0.447	23.61	1.3207	6.67	911	3.52	89	79
07/05/06	B20	1,166	Greg	2,880	29.78	0.474	26.55	1.5385	6.65	935	3.43	91	78
Average				2,895	30.26	0.454	24.49	1.4761	6.62	935	3.44	79	77
Standard Deviation				49	0.40	0.024	1.80	0.0936	0.10	15	0.05	9	2
Coefficient of Variation				1.7%	1.3%	5.3%	7.3%	6.3%	1.4%	1.6%	1.6%	11.3%	2.0%

Table 39. Freightliner Class 8 Truck – Freeway Cycle – 2007 Cert Diesel

Date	Fuel	Run	Driver	CO2 (g/mile)	NOx (g/mile)	THC (g/mile)	CO (g/mile)	PM (g/mile)	Distance (miles)	Fuel Cons (g/mile)	Fuel Econ (mpg)	Humidity (grain/lb)	Temp (F)
07/06/06	Base	1,171	Greg	1,699	22.25	0.216	8.09	0.4767	15.50	538	5.91	85	77
07/06/06	Base	1,172	Greg	1,745	22.54	0.209	8.17	0.6062	15.50	545	5.84	85	79
07/06/06	Base	1,173	Greg	1,741	22.08	0.172	7.93	0.5470	15.51	542	5.87	96	78
07/07/06	Base	1,183	Greg	1,708	22.11	0.211	8.78	0.4604	15.49	537	5.92	93	79
07/07/06	Base	1,184	Greg	1,689	21.65	0.190	8.19	0.4138	15.51	535	5.95	93	77
07/07/06	Base	1,185	Greg	1,688	22.98	0.201	7.68	0.3913	15.50	538	5.92	94	77
Average				1,712	22.27	0.200	8.14	0.4826	15.50	539	5.90	91	78
Standard Deviation				25	0.45	0.016	0.37	0.0813	0.01	4	0.04	5	1
Coefficient of Variation				1.5%	2.0%	8.2%	4.5%	16.8%	0.0%	0.7%	0.6%	5.3%	1.2%

Table 40. Freightliner Class 8 Truck – Freeway Cycle – B20 (2007 Cert Diesel and Agland Biodiesel)

Date	Fuel	Run	Driver	CO2 (g/mile)	NOx (g/mile)	THC (g/mile)	CO (g/mile)	PM (g/mile)	Distance (miles)	Fuel Cons (g/mile)	Fuel Econ (mpg)	Humidity (grain/lb)	Temp (F)
07/06/06	B20	1,176	Greg	1,732	23.31	0.134	7.80	0.4001	15.50	559	5.75	99	78
07/06/06	B20	1,177	Greg	1,720	23.04	0.168	7.64	0.3702	15.50	554	5.79	96	77
07/06/06	B20	1,178	Greg	1,717	23.04	0.166	7.63	0.3646	15.50	552	5.81	101	77
07/07/06	B20	1,180	Greg	1,721	23.69	0.182	7.41	0.3368	15.50	552	5.82	91	77
07/07/06	B20	1,181	Greg	1,713	23.10	0.173	7.55	0.3259	15.50	551	5.83	89	79
07/07/06	B20	1,182	Greg	1,695	22.27	0.184	7.43	0.3401	15.51	551	5.83	89	80
Average				1,716	23.08	0.168	7.58	0.3563	15.50	553	5.81	94	78
Standard Deviation				12	0.47	0.018	0.15	0.0274	0.00	3	0.03	5	1
Coefficient of Variation				0.7%	2.0%	10.8%	1.9%	7.7%	0.0%	0.5%	0.5%	5.4%	1.5%

Table 41. Motor Coach – CSHVC – 2007 Cert Diesel

Date	Fuel	Run	Driver	CO2 (g/mile)	NOx (g/mile)	THC (g/mile)	CO (g/mile)	PM (g/mile)	Distance (miles)	Fuel Cons (g/mile)	Fuel Econ (mpg)	Humidity (grain/lb)	Temp (F)
5/25/2006	Base	1,115	Greg	1,510	7.77	0.21	4.51	0.2686	6.67	487	6.60	71	72
5/25/2006	Base	1,116	Greg	1,492	7.87	0.23	4.41	0.2432	6.68	484	6.64	75	73
5/25/2006	Base	1,117	Greg	1,513	7.49	0.27	4.16	0.2926	6.68	489	6.58	75	73
5/26/2006	Base	1,127	Greg	1,504	7.78	0.22	4.06	0.2397	6.67	486	6.61	76	74
5/26/2006	Base	1,128	Greg	1,489	7.87	0.21	3.63	0.2345	6.68	481	6.68	68	74
5/26/2006	Base	1,129	Greg	1,501	7.70	0.23	3.56	0.2439	6.68	482	6.66	73	74
Average				1,501	7.75	0.23	4.05	0.2538	6.68	485	6.63	73	73
Standard Deviation				10	0.14	0.02	0.39	0.0224	0.01	3	0.04	3	1
Coefficient of Variation				0.6%	1.8%	10.3%	9.7%	8.8%	0.1%	0.6%	0.6%	4.1%	1.5%

Table 42. Motor Coach – CSHVC – B20 (2007 Cert Diesel and Agland Biodiesel)

Date	Fuel	Run	Driver	CO2 (g/mile)	NOx (g/mile)	THC (g/mile)	CO (g/mile)	PM (g/mile)	Distance (miles)	Fuel Cons (g/mile)	Fuel Econ (mpg)	Humidity (grain/lb)	Temp (F)
5/25/2006	B20	1,119	Greg	1,484	8.01	0.18	3.40	0.1797	6.68	491	6.59	71	75
5/25/2006	B20	1,120	Greg	1,488	8.23	0.19	3.30	0.1868	6.68	493	6.56	69	75
5/25/2006	B20	1,121	Greg	1,496	7.86	0.20	3.18	nm	6.67	497	6.52	73	75
5/26/2006	B20	1,123	Greg	1,495	7.80	0.20	2.88	0.1740	6.68	495	6.55	83	72
5/26/2006	B20	1,124	Greg	1,503	8.04	0.21	3.10	0.1810	6.67	493	6.57	69	72
5/26/2006	B20	1,125	Greg	1,507	7.82	0.20	3.03	0.1910	6.67	500	6.48	70	73
Average				1,496	7.96	0.19	3.15	0.1825	6.67	495	6.54	72	74
Standard Deviation				9	0.16	0.01	0.19	0.0066	0.01	3	0.04	5	2
Coefficient of Variation				0.6%	2.1%	4.5%	6.0%	3.6%	0.1%	0.7%	0.6%	7.5%	2.0%

Table 43. Motor Coach – UDDS – 2007 Cert Diesel

Date	Fuel	Run	Driver	CO2 (g/mile)	NOx (g/mile)	THC (g/mile)	CO (g/mile)	PM (g/mile)	Distance (miles)	Fuel Cons (g/mile)	Fuel Econ (mpg)	Humidity (grain/lb)	Temp (F)
5/31/2006	Base	1,142	Stuart	1,378	7.18	0.14	3.73	0.2355	5.53	443	7.23	69	78
5/31/2006	Base	1,143	Stuart	1,366	6.98	0.11	4.04	0.2478	5.52	442	7.26	75	76
5/31/2006	Base	1,144	Stuart	1,368	7.07	0.13	3.51	0.2231	5.54	441	7.26	76	75
6/2/2006	Base	1,146	Stuart	1,462	6.80	0.16	3.70	0.2480	5.53	470	6.83	78	71
6/2/2006	Base	1,147	Stuart	1,465	6.97	0.15	3.46	0.2448	5.53	470	6.84	75	72
6/2/2006	Base	1,148	Stuart	1,462	6.94	0.14	3.49	0.2330	5.53	469	6.85	73	71
Average				1,417	6.99	0.14	3.66	0.2387	5.53	456	7.05	74	74
Standard Deviation				51	0.13	0.02	0.22	0.0099	0.01	15	0.23	3	3
Coefficient of Variation				3.6%	1.9%	12.4%	6.0%	4.2%	0.1%	3.3%	3.2%	3.9%	4.0%

Table 44. Motor Coach – UDDS – B20 (2007 Cert Diesel and Agland Biodiesel)

Date	Fuel	Run	Driver	CO2 (g/mile)	NOx (g/mile)	THC (g/mile)	CO (g/mile)	PM (g/mile)	Distance (miles)	Fuel Cons (g/mile)	Fuel Econ (mpg)	Humidity (grain/lb)	Temp (F)
5/31/2006	B20	1,136	Stuart	1,390	7.58	0.09	2.82	0.1530	5.52	455	7.12	70	74
5/31/2006	B20	1,137	Stuart	1,401	7.44	0.14	2.93	0.1552	5.52	461	7.02	76	73
5/31/2006	B20	1,140	Stuart	1,445	7.12	0.13	2.85	0.1521	5.53	476	6.80	76	74
6/2/2006	B20	1,149	Stuart	1,418	7.13	0.15	2.99	0.1723	5.53	466	6.95	72	73
6/2/2006	B20	1,150	Stuart	1,406	6.95	0.14	3.04	0.1814	5.53	460	7.04	69	73
6/2/2006	B20	1,151	Stuart	1,399	7.14	0.15	3.09	0.1892	5.53	457	7.09	65	74
Average				1410	7.22	0.13	2.95	0.1672	5.53	462	7.00	71	73
Standard Deviation				20	0.24	0.02	0.11	0.0160	0.01	8	0.12	4	0
Coefficient of Variation				1.4%	3.3%	17.8%	3.6%	9.6%	0.1%	1.7%	1.7%	6.1%	0.6%

Table 45. Green Diesel School Bus – RUCSBC – 2007 Cert Diesel

Date	Fuel	Run	Driver	CO2 (g/mile)	NOx (g/mile)	THC (g/mile)	CO (g/mile)	PM (g/mile)	Distance (miles)	Fuel Cons (g/mile)	Fuel Econ (mpg)	Humidity (grain/lb)	Temp (F)
03/22/06	Base	993	Greg	2,134	8.78	0.03	0.04	0.0016	7.61	651	4.95	66	75
03/22/06	Base	994	Greg	2,149	nm	0.04	0.11	0.0019	7.61	654	4.92	58	76
03/22/06	Base	995	Greg	2,098	nm	0.02	0.06	0.0016	7.61	642	5.02	62	72
03/22/06	Base	996	Greg	2,091	nm	0.02	0.07	0.0013	7.62	644	5.00	67	70
03/23/06	Base	1,007	Greg	2,074	8.90	0.02	0.18	nm	7.63	632	5.10	65	74
03/23/06	Base	1,008	Greg	2,126	8.97	0.03	0.12	nm	7.62	651	4.94	64	74
03/23/06	Base	1,009	Greg	2,121	8.92	0.02	0.18	0.0015	7.62	652	4.93	66	74
03/23/06	Base	1,010	Greg	2,106	9.12	0.03	0.09	0.0015	7.64	651	4.94	63	74
03/24/06	Base	1,017	Greg	2,120	8.84	0.02	0.09	0.0013	7.61	648	4.97	66	74
03/24/06	Base	1,018	Greg	2,134	8.98	0.02	0.10	0.0008	7.62	654	4.92	66	74
Average				2,115	8.93	0.02	0.10	0.0014	7.62	648	4.97	65	74
Standard Deviation				23	0.11	0.01	0.05	0.0003	0.01	7	0.06	3	2
Coefficient of Variation				1.1%	1.2%	31.9%	45.5%	22.1%	0.1%	1.1%	1.1%	4.1%	2.2%

Table 46. Green Diesel School Bus – RUCSBC – B20 (2007 Cert Diesel and Agland Biodiesel)

Date	Fuel	Run	Driver	CO2 (g/mile)	NOx (g/mile)	THC (g/mile)	CO (g/mile)	PM (g/mile)	Distance (miles)	Fuel Cons (g/mile)	Fuel Econ (mpg)	Humidity (grain/lb)	Temp (F)
03/22/06	B20	998	Greg	2,098	8.83	0.05	0.09	0.0016	7.61	652	4.98	67	70
03/22/06	B20	999	Greg	2,101	nm	0.02	0.05	0.0019	7.62	647	5.02	64	70
03/22/06	B20	1,000	Greg	2,131	nm	0.01	0.03	0.0016	7.62	667	4.86	64	70
03/22/06	B20	1,001	Greg	2,110	9.14	0.02	0.03	nm	7.61	656	4.94	64	71
03/23/06	B20	1,003	Greg	2,127	9.34	0.02	0.04	0.0019	7.61	660	4.92	58	71
03/23/06	B20	1,004	Greg	2,124	9.21	0.02	0.06	0.0020	7.60	667	4.87	65	73
03/23/06	B20	1,005	Greg	2,120	9.18	0.02	0.12	0.0011	7.62	661	4.91	65	73

Table 46. (Continued).

Date	Fuel	Run	Driver	CO2 (g/mile)	NOx (g/mile)	THC (g/mile)	CO (g/mile)	PM (g/mile)	Distance (miles)	Fuel Cons (g/mile)	Fuel Econ (mpg)	Humidity (grain/lb)	Temp (F)
Average				2,116	9.14	0.02	0.06	0.0017	7.61	659	4.93	64	71
Standard Deviation				13	0.19	0.01	0.03	0.0003	0.01	7	0.06	3	1
Coefficient of Variation				0.6%	2.1%	55.0%	55.0%	19.6%	0.1%	1.1%	1.2%	4.2%	1.8%

Table 47. Green Diesel School Bus – CSHVC – 2007 Cert Diesel

Date	Fuel	Run	Driver	CO2 (g/mile)	NOx (g/mile)	THC (g/mile)	CO (g/mile)	PM (g/mile)	Distance (miles)	Fuel Cons (g/mile)	Fuel Econ (mpg)	Humidity (grain/lb)	Temp (F)
03/14/06	Base	950	Greg	1,734	7.99	0.03	0.35	0.0014	6.68	567	5.66	64	73
03/14/06	Base	951	Greg	1,700	7.79	0.04	0.09	0.0011	6.70	554	5.80	64	72
03/14/06	Base	952	Greg	1,719	7.38	0.05	0.14	0.0011	6.70	559	5.75	66	72
03/15/06	Base	954	Greg	1,716	7.67	-0.05	0.25	0.0012	6.69	555	5.80	65	71
03/15/06	Base	955	Greg	1,706	7.47	0.04	0.12	0.0009	6.70	565	5.70	67	76
03/15/06	Base	956	Greg	1,693	7.41	0.02	0.07	0.0016	6.70	556	5.78	67	75
03/16/06	Base	966	Greg	1,684	8.00	0.02	0.26	0.0009	6.71	536	6.01	65	72
03/16/06	Base	967	Greg	1,676	7.67	0.03	0.06	0.0008	6.71	539	5.97	69	72
03/16/06	Base	968	Greg	1,670	7.56	0.03	0.08	0.0006	6.71	545	5.91	59	73
03/21/06	Base	988	Greg	1,692	7.99	0.04	0.42	0.0007	6.69	544	5.91	63	73
03/21/06	Base	989	Greg	1,675	nm	-0.01	-0.02	0.0010	6.69	538	5.98	64	75
03/21/06	Base	990	Greg	1,684	nm	0.03	0.02	0.0003	6.69	544	5.91	65	73
03/21/06	Base	991	Greg	1,662	7.83	0.02	0.07	0.0006	6.69	540	5.96	65	71
Average				1,693	7.70	0.02	0.15	0.0009	6.70	549	5.86	65	73
Standard Deviation				21	0.23	0.03	0.13	0.0004	0.01	11	0.11	2	1
Coefficient of Variation				1.2%	3.0%	118.7%	90.1%	37.3%	0.1%	1.9%	1.9%	3.6%	1.9%

Table 48. Green Diesel School Bus – CSHVC – B20 (2007 Cert Diesel and Agland Biodiesel)

Date	Fuel	Run	Driver	CO2 (g/mile)	NOx (g/mile)	THC (g/mile)	CO (g/mile)	PM (g/mile)	Distance (miles)	Fuel Cons (g/mile)	Fuel Econ (mpg)	Humidity (grain/lb)	Temp (F)
03/15/06	B20	958	Greg	1,716	7.75	0.02	0.02	0.0011	6.70	568	5.71	72	71
03/15/06	B20	959	Greg	1,707	7.54	0.04	0.18	0.0014	6.70	560	5.78	65	70
03/15/06	B20	960	Greg	1,721	7.77	0.02	0.13	0.0010	6.69	554	5.85	69	67
03/16/06	B20	962	Greg	1,704	7.70	0.04	0.19	0.0012	6.70	574	5.65	57	74
03/16/06	B20	963	Greg	1,689	7.60	0.03	0.12	0.0014	6.70	566	5.72	64	73
03/16/06	B20	964	Greg	1,685	7.49	0.04	0.09	0.0013	6.70	566	5.73	64	73
Average				1,704	7.64	0.03	0.12	0.0012	6.70	565	5.74	65	71
Standard Deviation				14	0.12	0.01	0.06	0.0002	0.00	7	0.07	5	2
Coefficient of Variation				0.8%	1.5%	31.5%	52.2%	14.5%	0.1%	1.2%	1.2%	8.2%	3.2%

Table 49. Conventional School Bus – RUCSBC – 2007 Cert Diesel

Date	Fuel	Run	Driver	CO2 (g/mile)	NOx (g/mile)	THC (g/mile)	CO (g/mile)	PM (g/mile)	Distance (miles)	Fuel Cons (g/mile)	Fuel Econ (mpg)	Humidity (grain/lb)	Temp (F)
08/09/06	Base	1,228	John	1,984	10.10	0.265	8.03	0.6554	7.61	643	4.93	88	76
08/09/06	Base	1,229	John	1,935	9.94	0.266	7.57	0.6193	7.60	622	5.09	84	76
08/09/06	Base	1,230	John	1,937	9.79	0.284	8.31	0.7038	7.65	627	5.05	88	75
08/09/06	Base	1,231	John	1,944	9.94	0.278	8.24	0.6731	7.63	624	5.07	89	75
08/11/06	Base	1,242	John	1,983	9.63	0.402	8.80	0.7604	7.63	636	4.98	87	75
08/11/06	Base	1,243	John	1,990	9.66	0.727	10.10	0.7370	7.63	645	4.91	88	76
08/11/06	Base	1,244	John	1,987	9.75	0.483	9.27	0.7303	7.63	640	4.95	86	75
08/11/06	Base	1,245	John	1,940	9.81	0.526	9.75	0.6841	7.63	625	5.07	86	74
08/17/06	Base	1,268	John	1,998	9.81	0.259	7.97	nm	7.65	640	4.96	89	74
08/17/06	Base	1,269	John	1,989	9.38	0.357	10.36	nm	7.63	636	4.98	88	76
08/17/06	Base	1,270	John	1,968	nm	0.320	9.50	nm	7.62	626	5.06	86	77

Table 49. (Continued).

Date	Fuel	Run	Driver	CO2 (g/mile)	NOx (g/mile)	THC (g/mile)	CO (g/mile)	PM (g/mile)	Distance (miles)	Fuel Cons (g/mile)	Fuel Econ (mpg)	Humidity (grain/lb)	Temp (F)
08/17/06	Base	1,271	John	1,968	nm	0.298	9.72	nm	7.63	630	5.03	87	76
08/17/06	Base	1,272	John	1,973	nm	0.383	8.75	nm	7.62	634	5.00	93	74
Average				1,969	9.78	0.373	8.95	0.6954	7.63	633	5.01	88	75
Standard Deviation				22	0.20	0.137	0.90	0.0467	0.01	8	0.06	2	1
Coefficient of Variation				1.1%	2.0%	36.6%	10.0%	6.7%	0.2%	1.2%	1.2%	2.5%	1.3%

Table 50. Conventional School Bus – RUCSBC – B20 (2007 Cert Diesel and Agland Biodiesel)

Date	Fuel	Run	Driver	CO2 (g/mile)	NOx (g/mile)	THC (g/mile)	CO (g/mile)	PM (g/mile)	Distance (miles)	Fuel Cons (g/mile)	Fuel Econ (mpg)	Humidity (grain/lb)	Temp (F)
08/09/06	B20	1,224	John	1,999	10.78	0.179	5.09	0.4875	7.67	657	4.90	87	74
08/09/06	B20	1,225	John	1,968	10.57	0.185	5.68	0.4818	7.60	646	4.98	87	75
08/09/06	B20	1,226	John	1,965	10.28	0.197	5.80	0.4818	7.64	641	5.02	85	78
08/09/06	B20	1,227	John	1,991	10.41	0.206	6.63	0.5317	7.63	654	4.92	82	77
08/11/06	B20	1,247	John	1,928	10.22	0.416	7.85	0.5171	7.62	632	5.10	89	75
08/11/06	B20	1,248	John	1,949	10.35	0.459	8.40	0.5820	7.62	643	5.00	84	75
08/11/06	B20	1,249	John	1,954	10.12	0.455	9.05	0.6172	7.62	642	5.02	91	75
Average				1965	10.39	0.300	6.93	0.5284	7.63	645	4.99	86	76
Standard Deviation				24	0.22	0.135	1.52	0.0531	0.02	8	0.07	3	2
Coefficient of Variation				1.2%	2.2%	45.2%	21.9%	10.0%	0.3%	1.3%	1.3%	3.6%	2.0%

Table 51. Conventional School Bus – CSHVC – 2007 Cert Diesel

Date	Fuel	Run	Driver	CO2 (g/mile)	NOx (g/mile)	THC (g/mile)	CO (g/mile)	PM (g/mile)	Distance (miles)	Fuel Cons (g/mile)	Fuel Econ (mpg)	Humidity (grain/lb)	Temp (F)
08/01/06	Base	1,193	Greg	1,650	10.24	0.338	4.87	0.1954	6.69	533	5.95	96	74
08/01/06	Base	1,194	Greg	1,652	10.22	0.336	4.68	0.2016	6.67	544	5.83	80	75
08/01/06	Base	1,195	Greg	1,644	9.90	0.402	4.89	0.1882	6.69	538	5.90	81	78
08/02/06	Base	1,206	Greg	1,624	9.89	0.603	5.87	0.2114	6.69	543	5.83	77	78
08/02/06	Base	1,207	Greg	1,624	9.76	0.480	6.47	0.2517	6.70	537	5.89	83	79
08/03/06	Base	1,209	Greg	1,634	10.04	0.432	5.24	0.1972	6.70	535	5.93	81	73
08/03/06	Base	1,210	Greg	1,607	9.88	0.463	5.94	0.2371	6.69	528	6.01	86	74
08/03/06	Base	1,211	Greg	1,611	9.85	0.475	6.17	0.2439	6.70	517	6.13	85	74
08/03/06	Base	1,212	Greg	1,618	9.67	0.491	6.31	0.2429	6.70	531	5.97	97	77
08/10/06	Base	1,233	Greg	1,637	9.94	0.433	4.61	0.1436	6.69	535	5.93	86	74
08/10/06	Base	1,234	Greg	1,622	9.85	0.264	4.41	0.1460	6.68	528	6.00	85	73
08/10/06	Base	1,235	Greg	1,615	9.60	0.440	4.52	0.1417	6.69	535	5.92	84	73
08/10/06	Base	1,236	Greg	1,614	9.58	0.487	5.14	0.1624	6.69	539	5.88	84	74
08/10/06	Base	1,237	Greg	1,606	9.73	0.431	4.38	0.1370	6.70	537	5.90	84	73
08/10/06	Base	1,239	Greg	1,610	9.66	0.507	4.76	nm	6.70	537	5.90	85	73
Average				1,620	9.85	0.439	5.22	0.1929	6.69	534	5.93	85	75
Standard Deviation				11	0.20	0.081	0.74	0.0414	0.01	7	0.08	5	2
Coefficient of Variation				0.7%	2.0%	18.6%	14.1%	21.5%	0.1%	1.2%	1.3%	6.3%	2.8%

Table 52. Conventional School Bus – CSHVC – B20 (2007 Cert Diesel and Agland Biodiesel)

Date	Fuel	Run	Driver	CO2 (g/mile)	NOx (g/mile)	THC (g/mile)	CO (g/mile)	PM (g/mile)	Distance (miles)	Fuel Cons (g/mile)	Fuel Econ (mpg)	Humidity (grain/lb)	Temp (F)
08/01/06	B20	1,198	Greg	1,624	9.83	0.419	6.22	0.2211	6.69	554	5.81	102	77
08/01/06	B20	1,199	Greg	1,623	9.65	0.473	7.06	0.2494	6.68	554	5.80	104	78
08/01/06	B20	1,200	Greg	1,608	9.51	0.439	6.90	0.2464	6.69	551	5.83	104	78
08/02/06	B20	1,202	Greg	1,641	10.21	0.346	4.69	0.1685	6.69	544	5.91	77	75
08/02/06	B20	1,203	Greg	1,622	9.73	0.386	5.46	0.1978	6.70	538	5.97	77	75
08/02/06	B20	1,204	Greg	1,615	9.63	0.336	5.60	0.2004	6.69	546	5.89	77	75
08/02/06	B20	1,205	Greg	1,624	9.60	0.447	6.48	0.2260	6.69	548	5.87	78	76
08/03/06	B20	1,214	Greg	1,616	9.95	0.383	6.18	0.2328	6.69	545	5.90	78	80
08/03/06	B20	1,215	Greg	1,602	9.67	0.395	5.94	0.2266	6.69	543	5.92	77	76
08/03/06	B20	1,216	Greg	1,592	9.82	0.407	6.11	0.2218	6.69	538	5.98	102	75
08/04/06	B20	1,218	Greg	1,641	9.96	0.515	5.96	0.2029	6.69	544	5.92	75	79
08/04/06	B20	1,219	Greg	1,620	9.61	0.473	5.92	0.1856	6.68	548	5.87	93	74
08/04/06	B20	1,220	Greg	1,626	9.71	0.531	5.98	0.1938	6.69	554	5.80	76	77
08/04/06	B20	1,221	Greg	1,620	9.77	0.756	6.81	0.2119	6.69	556	5.78	91	74
08/04/06	B20	1,222	Greg	1,612	9.53	0.562	6.59	0.2125	6.69	553	5.82	92	75
08/15/06	B20	1,251	Greg	1,673	9.55	0.352	4.96	0.1724	6.69	553	5.83	90	75
08/15/06	B20	1,252	Greg	1,656	9.55	0.355	4.50	0.1647	6.69	549	5.87	88	79
08/15/06	B20	1,254	Greg	1,655	10.43	0.326	3.58	0.1096	6.69	555	5.81	86	80
08/15/06	B20	1,255	Greg	1,645	10.24	0.342	3.68	0.1113	6.69	555	5.81	86	81
Average				1,627	9.79	0.434	5.72	0.1977	6.69	549	5.86	87	77
Standard Deviation				20	0.26	0.105	1.01	0.0391	0.00	6	0.06	11	2
Coefficient of Variation				1.2%	2.7%	24.1%	17.7%	19.8%	0.1%	1.0%	1.0%	12.1%	2.9%

REFERENCES

[1] Dtn Energy's Alternative Fuels Index. Vol. 4, Issue 32, p. 1 (2006).

[2] Tyson, K.; Bozell, J.; Wallace, R. ; Petersen, E. ; Moens, L. "Biomass Oil Analysis: Research Needs and Recommendations", Technical Report, National Renewable Energy Laboratory/TP-510-34796, June 2004.

[3] Sheehan, J.; Camobreco, V. ; Duffield, J. ; Graboski, M. ; Shapouri, H. "An Overview of Biodiesel and Petroleum Diesel Life Cycles", National Renewable Energy Laboratory/TP-580-24772, May 1998.

[4] McCormick, R.; Alleman, T. "Impact of Biodiesel Fuel on Pollutant Emissions from Diesel Engines" in *The Biodiesel Handbook*, Knothe, G.; Van Gerpen, J.; Krahl, J., ed.s, AOCS Press, 2005.

[5] Liotta, F.; Montalvo, D. "The Effect of Oxygenated Fuels on Emissions from a Modern Heavy-Duty Diesel Engine", SAE Tech. Pap. No. 932734 (1993).

[6] Rantanen, L.; Mikkonen, S.; Nylund, L.; Kociba, P.; Lappi, M.; Nylund, N. "Effect of Fuel on the Regulated, Unregulated and Mutagenic Emissions of DI Diesel Engines", SAE Tech. Pap. No. 932686 (1993).

[7] Sharp, C. "Transient Emissions Testing of Biodiesel and Other Additives in a DDC Series 60 Engine", Southwest Research Institute, Final Report to the National Biodiesel Board, December 1994, available at www.biodiesel.org.

[8] Sharp, C. "Emissions and Lubricity Evaluation of Rapeseed Derived Biodiesel Fuels", Southwest Research Institute, Final Report for Montana Department of Environmental Quality, November 1996, available at www.biodiesel.

[9] Graboski, M.; Ross, J.; McCormick, R. "Transient Emissions from No. 2 Diesel and Biodiesel Blends in a DDC Series 60 Engine", SAE Tech. Pap. No. 961166 (1996).

[10] McCormick; R., Ross, J.; Graboski, M. "Effect of Several Oxygenates on Regulated Emissions from Heavy-Duty Diesel Engines", *Environmental Science & Technology* 31, 1144-1150 1997).

[11] Starr, M. "Influence on Transient Emissions at Various Injection Timings, Using Cetane Improvers, Bio-Diesel, and Low Aromatic Fuels", SAE Tech. Pap. No. 972904(1997).

[12] Clark, N.; Atkinson, C.; Thompson, G.; Nine, R. "Transient Emissions Comparisons of Alternative Compression Ignition Fuels", SAE Tech. Pap. No. 1999-01-1117 (1999).

[13] Sharp, C.; Howell, S.; Jobe, J. "The Effect of Biodiesel Fuels on transient Emissions from Modern Diesel Engines, Part I Regulated Emissions and Performance", SAE Tech. Pap. No. 2000-01-1967 (2000).

[14] McCormick, R.; Alleman, T.; Graboski, M.; Herring, A.; Tyson, K. "Impact Of Biodiesel Source Material and Chemical Structure On Emissions of Criteria Pollutants from a Heavy-Duty Engine", *Environmental Science & Technology* 35 1742-1747 (2001).

[15] McCormick, R.; Alvarez, J.; Graboski, M.; Tyson, K.; Vertin, K. " Fuel Additive and Blending Approaches to Reducing NOx Emissions from Biodiesel", SAE Tech. Pap. No. 2002-01-1658 (2002).

[16] United States Environmental Protection Agency. "A Comprehensive Analysis of Biodiesel Impacts on Exhaust Emissions", Draft Technical Report, EPA420-P-02 001, (2002).

[17] McGill, R.; Storey, J.; Wagner, R.; Irick, D.; Aakko, P.; Westerholm, M.; Nylund, N.; Lappi, M. "Emission Performance of Selected Biodiesel Fuels", SAE Tech. Pap. No. 2003-01-1866 (2003).

[18] Aakko, P.; Westerholm, M.; Nylund, N.O.; Moissio, M.; Marjamäki, M.; Mäkelä, T.; Hillamo, R. "Emission Performance Of Selected Biodiesel Fuels - VTT's Contribution", Research Report ENE5/33/2000, October 2000.

[19] Frank, B.; Tang, S.; Lanni, T.; Rideout, G.; Beregszaszy, C.; Meyer, N.; Chatterjee, S.; Conway, R.; Lowell, D.; Bush, C.; Evans, J. "A Study of the Effects of Fuel Type and Emission Control Systems on Regulated Gaseous Emissions from Heavy-Duty Diesel Engines", SAE Tech. Pap. No. 2004-01-1085 (2004).

[20] Souligny, M., Graham, L., Rideout, G.; Hosatte, P. "Heavy Duty Diesel Engine Performance and Comparative Emission Measurements for Different Biodiesel Blends Used in the Montreal BIOBUS Project", SAE Tech. Pap. No. 2004-01-1861 (2004).

[21] Alam, M.; Song, J.; Acharya, R.; Boehman, A.; Miller, K. "Combustion and Emissions Performance of Low Sulfur, Ultra Low Sulfur and Biodiesel Blends in a DI Diesel Engine", SAE Tech. Pap. No. 2004-01-3024 (2004).

[22] McCormick, R.; Tennant, C.; Hayes, R.; Black, S.; Ireland, J.; McDaniel, T.; Williams, A.; Frailey, M.; Sharp, C. "Regulated Emissions from Biodiesel Tested in Heavy-Duty Engines Meeting 2004 Emission Standards", SAE Tech. Pap. No. 2005 01-2200 (2005).

[23] Environment Canada. "Emissions Characterization of a Caterpillar 3126E Equipped with a Prototype SCRT System with Ultra Low Sulphur Diesel and a Biodiesel Blend", ERMD Report # 2005-32 (2005).

[24] Clark, N.; Lyons, D. "Class 8 Truck Emissions Testing: Effects of Test Cycles and Data on Biodiesel Operation", Transactions of the ASAE 42 (5) 1211-1219 (1999).

[25] Wang, W.; Lyons, D.; Clark, N.; Gautam, M.; Norton, P. "Emissions from Nine Heavy Trucks Fueled by Diesel and Biodiesel Blend without Engine Modification", Environmental Science & Technology 34 933-939 (2000).

[26] Peterson, C.; Taberski, J.; Thompson, J.; Chase, C.L. "The Effect of Biodiesel Feedstock on Regulated Emissions in Chassis Dynamometer Tests of a Pickup Truck", Transactions of the ASAE 43 (6) 1371-1381 (2000).

[27] Durbin, T.; Cocker, K.; Collins, J.; Norbeck , J.. "Evaluation of the Effects of Biodiesel and Biodiesel Blends on Exhaust Emission Rates and Reactivity – 2", Final Report from the Center for Environmental Research and Technology, University of California, August 2001.

[28] Environment Canada. "Emission Testing for NRCan Biodiesel Byway Pilot Project", private communication, May 2005.

[29] Holden, B.; Jack, J.; Miller, W.; Durbin, T. "Effect of Biodiesel on Diesel Engine Nitrogen Oxide and Other Regulated Emissions", Technical Report TR-2275-ENV, Naval Facilities Engineering Command, Port Hueneme, California. May 2006.

[30] United States Environmental Protection Agency. "Final Rule on In-Use Testing Program for Heavy-Duty Diesel Engines and Vehicles", EPA420-F-05-021, June 2005.

[31] Frey, H.; Kim, K. "Operational Evaluation of Emissions and Fuel Use of B20 versus Diesel Fueled Dump Trucks", Research Project No. 2004-18 FHWA/NC/2005-07, September 2005.

[32] Hearne, J.; Toback, A.; Akers, J.; Hesketh, R.; Marchese, A. "Development of a New Composite School Bus Test Cycle and the Effect of Fuel Type on Mobile Emissions from Three School Buses", SAE Tech. Pap. No. 2005-01-1616 (2005).

[33] Farzeneh, M.; Zietsman, J.; Perkinson, D. "School Bus Biodiesel (B20) NO_x Emissions Testing", Texas Transportation Institute, August 2006.

[34] United States Environmental Protection Agency. Engine Certification Information Center: http://www.epa.gov/oms/certdata.htm#largeng, accessed September 19, 2006.

[35] United States Census Bureau. "2002 Economic Census: Vehicle Inventory and Use Survey", EC02TV-US, December 2004.

[36] Zhang, Y.; Van Gerpen, J. "Combustion Analysis of Esters of Soybean Oil in a Diesel Engine" SAE Tech. Pap. No. 960765 (1996).

[37] United States Environmental Protection Agency. "The Effect of Cetane Number Increase Due to Additives on NOx Emissions from Heavy-Duty Highway Engines", Final Technical Report, EPA420-R-03-002, February 2003.

[38] Cheng, A.; Upatnieks, A.; Mueller, C. "Investigation of the Impact of Biodiesel Fuelling On NO_x Emissions Using an Optical Direct Injection Diesel Engine", to be published: *International Journal of Engine Research* 7 (4) 297-319 2006.

[39] Ban-Weiss, G.; Chen, J.; Buchholz, B.; Dibble, R. "A Numerical Investigation into the Anomalous Slight NOx Increase When Burning Biodiesel: A New (Old) Theory", *Prepr. Pap.-Am. Chem. Soc., Div. Fuel Chem.* 51(1), 24 (2006)

[40] Miller, J.; Bowman, C. "Mechanism and Modeling of Nitrogen Chemistry in Combustion", *Prog. Energy Combust. Sci.* 15 287 (1989).

[41] Hess, M.; Haas, M.; Foglia, T.; Marmer, W. "Effect of Antioxidant Addition on NO_x Emissions from Biodiesel", *Energy & Fuels* 19 1749 (2005).

[42] Tat, M.; van Gerpen, J. "Measurement of Biodiesel Speed of Sound and Its Impact on Injection Timing", National Renewable Energy Laboratory, NREL/SR-510-31462, February 2003.

[43] Monyem, A.; van Gerpen, J.; Canakci, M. "The Effect of Timing and Oxidation on Emissions from Biodiesel-Fueled Engines", *Trans. of the Am. Soc. of Agricultural Engineers* 44 35 (2001).

[44] Sybist, J.; Boehman, A. "Behavior Of A Diesel Injection System With Biodiesel Fuel", SAE Technical Paper No. 2003-01-1039 (2003).

[45] Clark, N.; Daley, J.; Nine, R.; Atkinson, C.M. "Application of the New City-Suburban Heavy Vehicle Route (CSHVR) to Truck Emissions Characterization", SAE Tech. Pap. No. 1999-01-1467 (1999).

[46] Code of Federal Regulations. Vol. 40, Subpart M.

[47] Zou, Z.; Davis, S.; Beaty, K.; O'Keefe, M.; Hendricks, T.; Rehn, R.; Weissner, S.; Sharma, V.K. "A New Composite Drive Cycle for Heavy-Duty Hybrid Electric Class 4-6 Vehicles", SAE Tech. Pap. No. 2004-01-1052 (2004).

[48] Proc, K.; Barnitt, R.; Hayes, R.; McCormick, R.; Ha, L.; Fang, H. "100,000 Mile Evaluation of Transit Buses Operated on Biodiesel Blends (B20)", SAE Tech. Pap. No. 2006-01-3253 (2006).

[49] Yanowitz, J.; Graboski, M.; McCormick, R. "On the Prediction of In-Use Emissions of Heavy-Duty Diesel Vehicles from Engine Testing", *Environmental Science & Technology* 36 270 (2002).

In: Biodiesel Fuels Reexamined
Editor: Bryce A. Kohler

ISBN: 978-1-60876-140-1
© 2011 Nova Science Publishers, Inc.

Chapter 3

LIFE-CYCLE ASSESSMENT OF ENERGY AND GREENHOUSE GAS EFFECTS OF SOYBEAN-DERIVED BIODIESEL AND RENEWABLE FUELS[*]

H. Huo, M. Wang, C. Bloyd and V. Putsche

ACRONYMS AND ABBREVIATIONS

BFW	boiler feed water
BP	British Petroleum
CETC	CANMET Energy Technology Centre
CH_4	methane
CIDI	compression-ignition, direct-injection
CO	carbon monoxide
CO_2	carbon dioxide
CSO	clarified slurry oil
$DeCO_2$	decarboxylation
DOE	U.S. Department of Energy
ERS	Economic Research Service (USDA)
FCC	fluidized catalytic cracker
GHG	greenhouse gas
GREET	Greenhouse Gases, Regulated Emissions, and Energy Use in Transportation
HDO	hydrodeoxygenation
IPCC	Intergovernmental Panel on Climate Change
K	potassium
LCA	life-cycle analysis
LCO	light-cycle oil
LHV	lower heating value
LPG	liquefied petroleum gas
LSD	low-sulfur diesel

---

[*] This is an edited, reformatted and augmented version of a Energy Systems Division's publication, dated March 12, 2008.

N	nitrogen
N2O	nitrous oxide
NaOH	sodium hydroxide
NG	natural gas
NO$_x$	nitrogen oxide
NRCan	Natural Resources Canada
NREL	National Renewable Energy Laboratory
P	phosphorus
PM10	particulate matter with a diameter of 10 micrometers or less
PM2.5	particulate matter with a diameter of 2.5 micrometers or less
PTW	pump-to-wheels
RFG	reformulated gasoline
SI	spark-ignition
SMR	steam methane reforming
SO$_x$	sulfur oxides
USDA	U.S. Department of Agriculture
VGO	vacuum gas oil
VOC	volatile organic compound
WTP	well-to-pump
WTW	well-to-wheels
WWT	wastewater treatment

Units of Measure

bpd	barrel(s) per day
Btu	British thermal unit(s)
bu	bushel(s)
C	degree(s) Celsius
F	degree(s) Fahrenheit
ft^3	cubic foot (feet)
g	gram(s)
gal	gallon(s)
h	hour(s)
ha	hectare(s)
kW	kilowatt(s)
kWh	kilowatt-hour(s)
L	liter(s)
lb	pound(s)
mmBTU	million Btu
ppm	part(s) per million
psia	pound(s) per square inch absolute
psig	pound(s) per square inch gauge
scf	standard cubic foot (feet)
USD	U.S. dollar(s)
yr	year(s)

ACKNOWLEDGMENTS

This work was sponsored by DOE's Office of Energy Efficiency and Renewable Energy. Argonne National Laboratory is a DOE laboratory managed by UChicago Argonne, LLC, under Contract No. DE-AC02-06CH11357.

We are grateful to our DOE sponsor, Linda Bluestein, for her support and input to this study. We also wish to thank Robert McCormick and Caley Johnson of the National Renewable Energy Laboratory for providing ASPEN simulation results and insights to this study. We thank Philip Heirigs of Chevron and Leland Tong of the National Biodiesel Board for providing helpful comments on an earlier draft version of this report. However, we are solely responsible for the content of this report.

ABSTRACT

We assessed the life-cycle energy and greenhouse gas (GHG) emission impacts of the following three soybean-derived fuels by expanding, updating, and using Argonne National Laboratory's Greenhouse Gases, Regulated Emissions, and Energy Use in Transportation (GREET) model: (1) biodiesel produced from soy oil transesterification, (2) renewable diesel produced from hydrogenation of soy oil by using two processes (renewable diesel I and II), and (3) renewable gasoline produced from catalytic cracking of soy oil. We used four allocation approaches to address the co-products: a displacement approach; two allocation methods, one based on energy value and one based on market value; and a hybrid approach that integrates both the displacement and allocation methods. Each of the four allocation approaches generates different results. The displacement method shows a 6–25% reduction in total energy use for the soybean-based fuels compared with petroleum fuels, except for renewable diesel II. The allocation and hybrid approaches show a 13–31% increase in total energy use. All soybean-derived fuels achieve a significant reduction (52–107%) in fossil energy use and in petroleum use (more than 85%). With the displacement approach, all four soybean-based fuels achieve modest to significant reductions (64–174%) in well-to-wheels GHG emissions. With the allocation and hybrid approaches, the fuels achieve a modest reduction in GHG emissions (57–74%). These results demonstrate the importance of the methods that are used in dealing with co-product issues for these renewable fuels.

1. INTRODUCTION

There has long been a desire to find alternative liquid fuel replacements for petroleum-based transportation fuels. Biodiesel, produced from seed oils or animal fats via the transesterification process, has been the focus of biofuel production because of its potential environmental benefits and because it is made from renewable biomass resources. Biodiesel can be derived from various biological sources such as seed oils (e.g., soybeans, rapeseeds, sunflower seeds, palm oil, jatropha seeds, waste cooking oil) and animal fats. In the United States, a majority of biodiesel is produced from soybean oil. In Europe (especially in Germany), biodiesel is produced primarily from rapeseeds. Biodiesel can be blended with conventional diesel fuel in any proportion and used in diesel engines without significant engine modifications (Keller et al. 2007). In recent years, the sales volume for biodiesel in the United States has increased dramatically: from about 2 million gallons in 2000, to 75 million gallons in 2005, to 250 million gallons in 2006 (National Biodiesel Board 2007).

Transesterification of seed oils and animal fats has been the major technology for biodiesel production to date. New process technologies based on hydrogenation to convert seed oils and animal fats to diesel fuel and gasoline have recently emerged. The CANMET Energy Technology Centre (CETC) of Natural Resources Canada (NRCan) has developed a technology to convert seed oils and animal fats into a high-cetane, low-sulfur diesel fuel blending stock called "SuperCetane" [(S&T)[2] Consultants Inc. 2004]. UOP developed conversion processes based on conventional hydroprocessing technologies that are already widely deployed in petroleum refineries. The hydro-generation technologies utilize seed oils or animal fats to produce an isoparaffin-rich diesel substitute referred to as "green diesel" (Kalnes et al. 2007). UOP also proposed a technology that can produce "green gasoline" by cracking seed oils and grease in a fluidized catalytic cracker (FCC) unit (UOP 2005). The diesel and gasoline produced from these processes are often referred to as renewable diesel and gasoline.

In this report, we present a life-cycle analysis of the energy and GHG emission impacts of biodiesel, renewable diesel, and renewable gasoline relative to those of petroleum diesel and gasoline. In the United States, soybeans are the major feedstock for biodiesel production now and, potentially, for renewable diesel and gasoline production in the future. In our study, we evaluated production of biodiesel, renewable diesel, and renewable gasoline from soybeans.

For this study, we expanded and updated the GREET (Greenhouse Gases, Regulated Emissions, and Energy Use in Transportation) model (see http://www. transportation.anl. gov/software/ GREET/index.html). In 1995, with funding from the U.S. Department of Energy (DOE), Argonne National Laboratory's Center for Transportation Research developed the GREET model for use in estimating the full fuel-cycle energy and emissions impacts of alternative transportation fuels and advanced vehicle technologies. Since that time, the model has been updated to include new fuels and transportation technologies. The latest version — GREET 1.8a — is capable of analyzing more than 100 transportation fuel pathways.

For a given vehicle and fuel system, GREET evaluates total energy use, fossil fuels, natural gas (NG) use, coal use, and petroleum use; emissions of carbon dioxide (CO_2)-equivalent greenhouse gases (GHGs) including CO_2, methane (CH_4), and nitrous oxide (N_2O); and emissions of six criteria pollutants — volatile organic compounds (VOCs), carbon monoxide (CO), nitrogen oxides (NO_x), particulate matter with a diameter of 10 micrometers or less (PM_{10}) and 2.5 micrometers or less ($PM_{2.5}$), and sulfur oxides (SO_x). These criteria pollutant emissions are further separated into total and urban emissions to reflect human exposure to air pollution caused by emissions of the six criteria pollutants.

Our analysis in this study includes the following six fuel pathways:

1) Conventional petroleum-based reformulated gasoline (RFG);
2) Conventional petroleum-based low-sulfur diesel (LSD) with 15 parts per million (ppm) sulfur content;
3) Soybean-based biodiesel produced by using the transesterification process;

4) Soybean-based renewable diesel I ("SuperCetane") produced by using the hydrogenation process;

5) Soybean-based renewable diesel II ("green diesel") produced by using the hydrogenation process; and

6) Renewable gasoline ("green gasoline") produced by using catalytic cracking.

We used petroleum gasoline and diesel as the baseline fuels; our analysis was conducted for year 2010. We estimated consumption of total energy, fossil energy, and petroleum oil and emissions of GHGs (CO_2, N_2O, and CH_4) for each of the six pathways. Figure 1-1 illustrates the system boundary for the six fuel pathways. The four soybean-based pathways consist of six stages: (1) farming activities, including manufacture of fertilizer and other chemicals, soybean farming, and soybean harvest; (2) soybean transportation from farms to processing plants; (3) soy oil extraction in processing plants; (4) production of biodiesel or other renewable fuels in plants; (5) fuel transportation and distribution from plants to refueling stations; and (6) fuel use during vehicle operation. As shown, the four soybean-based fuel pathways have three common stages: soybean farming, soybean transportation, and soy oil extraction. The four paths differ in terms of their fuel production processes and vehicle operations.

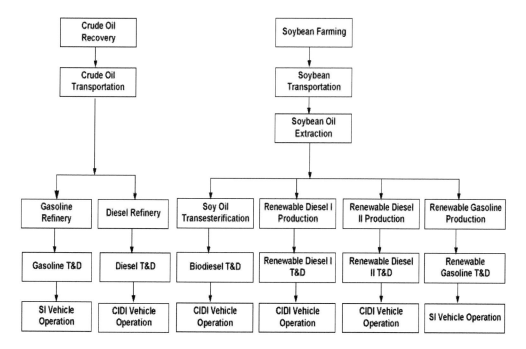

Figure 1-1. System Boundaries for Life-Cycle Analysis of Petroleum Gasoline and Diesel Fuels and Soybean-Based Biodiesel and Renewable Fuels.

The pathways for petroleum gasoline, petroleum diesel, and soybean-based biodiesel had been incorporated into the GREET model before this study. However, for this study, we updated soybean farming simulations in GREET with the latest U.S. Department of Agriculture data on energy and fertilizer use associated with soybean farming (USDA 2007a, b). We updated N_2O emission simulations for soybean fields by using newly released data from the Intergovernmental Panel on Climate Change (IPCC 2006). Moreover, we expanded GREET to include pathways for soybean-based renewable diesel and gasoline.

Process energy and mass balance data for the four soybean-based fuels are from our evaluation of available literature and process simulations by the National Renewable Energy Laboratory (NREL) using the ASPEN model. The processing of energy and mass balance data is described in Section 2. Section 3 presents the key issues regarding life-cycle simulations, gives GREET input assumptions, and compares the different production processes and fuel properties of soybean-derived fuels. Section 4 presents the approaches used to address co-product credits. Section 5 provides an analysis and comparison of the life-cycle (or well-to-wheels [WTW]) energy and emission results for the six pathways examined in this study. Section 6 presents our conclusions. Finally, Appendices 1 and 2 present ASPEN simulations by NREL.

Note that this study does not consider potential land use changes. Increased CO_2 emissions from potential land use changes are an input option in GREET, but it was not used in the current analysis since reliable data on potential land use changes induced by soybean-based fuel production are not available. Furthermore, the main objective of this study is to concentrate on the process related issues described above.

2. Production Processes of Soybean-Based Renewable Fuels

This section describes the three basic processes that have been proposed for renewable diesel and gasoline production: two for renewable diesel fuel and one for renewable gasoline. It also presents the results of the process modeling work undertaken by NREL to characterize the mass and energy balances associated with the three processes. The NREL-simulated results were inputs to the life-cycle analysis (LCA) described in Sections 3 and 4.

Table 2-1 provides a list of current and planned renewable energy diesel facilities. For example, ConocoPhillips is currently operating a 1,000-barrel-per-day (bpd) facility in Ireland using soybean and other vegetable oils; the company entered into a partnership with Tyson foods in April 2007 to produce up to 12,000 bpd from animal fat generated in the United States.

Refinery-based biofuels have received strong support from vehicle manufacturers, both in the United States and abroad, because their physical and chemical properties are similar to conventional petroleum-based fuels. Refinery-based biofuels have also been supported by major international oil companies because they can be delivered by using the existing fuel delivery infrastructure with no modifications.

Feedstocks that can be used in biofuel production processes include seed oils (e.g., soy, corn, canola, or palm oil), recycled oils (e.g., yellow grease or brown [trap] grease), and animal fats (e.g., tallow, lard, or fish oil). Table 2-2 lists current estimates of these oils, which amount to about 100,000 bpd (UOP 2005). Vegetable oils, particularly soybean-derived oils, are of particular interest in this study because (1) soy oil is the principal feedstock used in the United States for production of biodiesel via the transesterification process and (2) soy oil is a currently modeled pathway in GREET.

Table 2-1. Current and Planned Renewable Diesel Facilities

Company	Size (bpd)	Location	Online Date
ConocoPhillips	1,000	Ireland	2006
ConocoPhillips	12,000	United States	To be determined
British Petroleum (BP)	1,900	Australia	2007
Neste	3,400	Finland	2007
Neste	3,400	Finland	2009
Petrobras	4 × 4,000	Brazil	2007
UOP/Eni	6,500	Italy	2009

Table 2-2. Feedstock Availability for Renewable Diesel Production in the United States (UOP 2005)

Feedstock	Feedstock	Total U.S. Production (bpd)	Available for Conversion to Fuels (bpd)
Vegetable oils	Soybeans, corn, canola, palm	194,000	33,500
Recycled products	Yellow grease, brown (trap) grease	51,700	33,800
Animal fats	Tallow, lard, fish oil	71,000	32,500

Because crude oil and bio-feedstocks are derived from the same sources (i.e., crude oil owes its existence to plants and animals that have decomposed over 600 million years), the question arises: Why not add the bio-feedstocks directly to the feeds for conventional refineries? The answer is that the molecular structures of all of the bio-feedstocks listed in Table 2-2 contain significant amounts of oxygen that must be removed prior to their processing with other petroleum-based feedstocks. The two standard processes to remove oxygen from hydrocarbon feeds are hydrodeoxygenation (HDO) and decarboxylation (DeCO2). Under the proper conditions and with the addition of hydrogen, the HDO reaction, given in Equation 2-1, converts the oxygen in the product feed into plain water.

$$C_nCOOH + 3H_2 \rightarrow C_{n+1} + 2H_2O \tag{2-1}$$

In the DeCO2 reaction, shown in Equation 2-2, the oxygen in the feed is removed as simple CO_2 in a lead/hydrogen catalytic reaction.

$$\begin{array}{c} Pb/H \\ C_nCOOH \rightarrow C_n + CO_2 \end{array} \tag{2-2}$$

In reality, it is difficult to have a processing vessel where only one process occurs; in all the current renewable diesel design schemes, both reactions take place. The particular operating designs and conditions determine which process is favored. A basic tradeoff is that, in order to optimize the HDO reaction shown in Equation 2-1, additional hydrogen is required; production of the hydrogen can be expensive and can result in environmental impacts. On the other hand, the only byproduct of the HDO process (Equation 2-1) is water, while the principal by-product of the DeCO2-process (Equation 2-2) is CO_2 — a GHG that is of concern in life-cycle modeling. However, the CO_2 from this process is the CO_2 uptaken during soybean growth.

2.1. Renewable Diesel Production Based on SuperCetane

The first renewable diesel production pathway, renewable diesel I, was modeled after a process called SuperCetane that was originally developed in the 1980s at the Saskatchewan Research Council and is now being developed by NRCan's CETC.

The SupereCetane process is based on adapting a conventional hydrotreating process so it can operate under proprietary operating conditions. Figure 2-1 shows a general process schematic for the SuperCetane process. A number of reactions occur in the process, including hydrocracking, hydrotreating, and hydrogenation. The hydrocracking process breaks apart large molecules; the hydrotreating removes oxygen. The process uses a conventional commercial refinery hydrotreating catalyst and hydrogen to produce a hydrocarbon liquid. This liquid can be distilled into three basic fractions: naphtha, middle distillate (or SuperCetane), and waxy residues. The principal product, the middle distillate, can be produced at yields of 70–80%. Because of the high cetane number (around 100), CETC believes that SuperCetane may prove most valuable as a blending agent for lower-quality diesels (CETC undated).

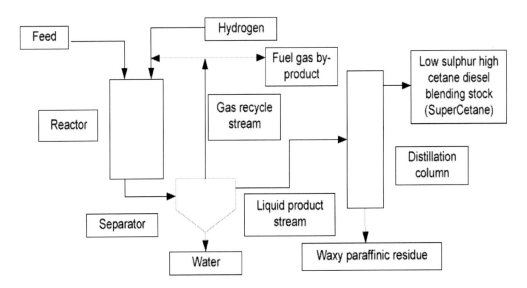

Figure 2-1. SuperCetane Process Flow (NRCan 2003).

The process has been used successfully in a 1-bpd pilot reactor. Feedstocks used in the pilot process include canola oil, soy oil, yellow grease, animal tallow, and tall oil (a by-product of the kraft pulping process). An important characteristic of this processing scheme is that internally generated fuel gas is combusted on site to meet facility steam requirements. Thus, all energy demands except electricity are met on site.

2.2. Renewable Diesel Production Based on UOP Hydrogenation Technology

The second renewable diesel production pathway, renewable diesel II, was modeled on a hydrogenation process developed by UOP, a leading supplier and licensor of process technology, catalysts, adsorbents, process plants, and consulting services to the petroleum refining, petrochemical, and gas processing industries. UOP, located in Des Plaines, Illinois, is a wholly owned subsidiary of Honeywell International. In 2005, UOP conducted a study for DOE entitled *Opportunities for Biorenewables in Oil Refineries* (UOP 2005). In November 2006, UOP announced the formation of a new Renewable Energy and Chemicals business unit focused on using the company's refinery skills to develop profitable and efficient ways to enable refineries to convert bio-feedstocks (e.g., vegetable oils and greases) into valuable fuels and chemicals.

UOP took another major step in June 2007, when the company announced that it had entered into an agreement with Eni S.p.A, a large European refiner, to build a 6,500-bpd renewable diesel unit in Livorno, Italy. The facility, which will process soy, rapeseed, palm, and other oils, is expected to come online in 2009. Facility operations will be based on a newly branded UOP process called Ecofining^{TM}. UOP has also announced that the technology that it developed in partnership with Eni integrates seamlessly into existing refinery operations and is currently available for licensing. The most recent license was granted to Galp Energia, Portugal's largest refiner, to develop a 6,500-bpd facility in Sines, Portugal (Reuters News 2007).

In its study for DOE, UOP examined two potential approaches for renewable diesel production. The first involved co-processing the bio-feedstock in an existing hydroprocessing unit; the second involved processing the bio-feedstock in a standalone processing unit. In order to design a process comparable to the CETC process modeled for renewable diesel I, the UOP standalone process scheme was characterized for this project by using ASPEN modeling. Figure 2-2 shows the basic production scheme for the UOP process in standalone mode.

In the standalone process, the bio-feedstock is fed into a diesel hydrotreater, where hydrogen and steam are added. An advantage of the UOP operating scheme is that, although the principal product is renewable diesel, the by-product is a valuable propane fuel mix. UOP reports that its resultant renewable diesel has a cetane value in the 70–90 range, offering significant blending benefits for existing refinery operations. UOP notes that when the standalone process is used, additional pretreatment is required to remove contaminants such as water, alkali metals, phosphorous, and ash. These would be removed by using a combination of existing equipment, such as hydrocyclones, desalting, acid washing, ion exchange, or fixed-guard bed catalyst systems (UOP 2005).

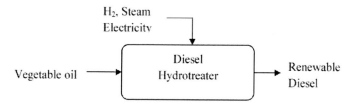

Figure 2-2. UOP-Proposed Standalone Renewable Diesel Production (UOP 2005).

2.3. Renewable Gasoline Production Based on UOP FCC Technology

As mentioned earlier, because bio-feedstocks are basically chains of carbon and hydrogen with added oxygen, standard refinery vessels could be modified to produce gasoline from these feedstocks. UOP has proposed such a scheme based on the use of an FCC unit (UOP 2005). (It should be noted that renewable gasoline is not nearly as far along the commercialization path as the renewable diesel processes discussed in Sections 2.2 and 2.3.) Figure 2-3 shows the general flow of the system proposed by UOP. As in the case of renewable diesel, the first step is pretreatment of the bio-feedstock; in this case, primarily to remove metals like calcium and potassium that would poison the FCC catalyst. Pretreatment also prevents metallurgy issues in the feed system, especially when processing greases. The pretreated oil is fed into the FCC unit along with the vacuum gas oil (VGO) stream. It should be noted that in the ASPEN modeling runs used to characterize renewable gasoline in Table 2-3, the FCC unit was characterized with only soybean oil feedstock. Although the standalone production of green gasoline would probably not be as economical as dual processing with VGO, it does allow for comparable life-cycle analysis, which is the principal thrust of this study. One of the differences between the renewable gasoline and the renewable diesel processes is that additional hydrogen is not required for the gasoline process. Another difference is that a significant portion of the energy value of the feedstock is contained in process by-products rather than the desired end product: renewable gasoline. The other principal product streams include light ends, light-cycle oil (LCO), and clarified slurry oil (CSO).

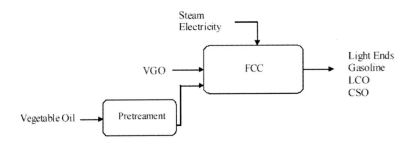

Figure 2-3. UOP Proposed Renewable Gasoline Production (UOP 2005).

2.4. ASPEN Model Results

A specific goal of the GREET WTW modeling has been to compare various transport fuels on a consistent basis. Consistency is achieved by basing model calculations on process mass and energy balances that are validated by using data from commercial operating facilities. Modeling of new renewable energy fuels thus presents a problem because facility mass and energy balances are either unavailable or available only from limited pilot plant operations that may not reflect mature commercial operating conditions.

For the three new fuels characterized in this report (pathways 4 through 6), NREL developed initial mass and energy balances by using the ASPEN process simulation model. The NRELmodeled mass and energy balances for the three fuels are listed in Table 2-3. Details of NREL's ASPEN simulations are presented in Appendices 1 and 2. Note that all data have been normalized to the basis of one pound of final fuel product. This adjustment allows the data to be incorporated into GREET on a consistent basis with existing fuel paths. The emissions presented in the table were estimated by using standard AP-42 emission factors.

Table 2-3. NREL-Simulated Renewable Fuels Mass and Energy Balances

Inputs and Outputs	Fuel		
	Renewable Diesel I (SuperCetane)	Renewable Diesel II (UOP-HDO)	Renewable Gasoline
Inputs (lb per lb of final fuel product)			
Soybean oil	1.510	1.174	2.2313
Hydrogen	0.030	0.032	
Steam		0.0329	0.0286
Air	0.9588		1.6782
Boiler feed water (BFW)			1.47
Outputs (lb/lb soybean oil)			
Renewable diesel	1.000	1.000	
Renewable gasoline			1.000
Fuel gas	0.253		
Product gas			0.3447
Heavies	0.175		
Water vapor	0.200		0.0287
Propane fuel mix		0.059	
CO2[a]	0.049	0.082	0.4103
LCO			0.2454
CSO			0.2914
Water-to-wastewater treatment (WWT)	0.0663	0.0971	0.2599
Return BFW/steam			1.47
O2	0.0201		0.0593
N2	0.7355		1.2675
Energy Inputs (unit per lb of final fuel product)			
Steam (Btu)	Process is self-sufficient in energy	84.05	−1,237
Electricity (kWh)	0.0394	0.0275	0.0544
CW (lb/h)	65.06	27.11	50.3

[a] This is the amount of CO2 from feedstock oil, which is eventually from the air during soybean growth.

To conduct the GREET analysis by using the three new renewable fuel pathways, additional component energy data are needed. The values used in the simulation were provided by NREL and are listed in Table 2-4. As data from commercial facility operations become available, the information will need to be updated to reflect any changes that might occur as the technologies mature.

The ASPEN simulations showed the mass and energy flow differences that were expected from proposed technology design schemes. For example, when renewable diesel I and renewable diesel II are compared, differences in hydrogen requirements, as well as the resultant CO_2 emissions, demonstrate the extent to which the HDO or $DeCO_2$ reaction was favored by the process design. Another difference is that all facility energy demands (except electricity) are met by recycling process-generated fuel gas in the renewable diesel I scheme. This process characteristic increases facility emissions and reduces facility energy by-products. These types of tradeoffs are central to the use of GREET in linking the new fuels to the existing fuel pathways in order to assess their life-cycle energy and GHG emission impacts

Table 2-4. NREL-Provided Base Energy Valuesof Renewable Fuel Components

Component	Lower Heating Value (Btu/lb)
Soybean oil	16,000
H2[a]	52,226
Renewable diesel I – SuperCetane	18,746
Renewable diesel II – UOP	18,925
Renewable gasoline	18,679
Fuel gas	27,999
Product gas	18,316
Heavies	20,617
Propane fuel mix	18,568
LCO	19,305
CSO	18,738

[a] Simulation of hydrogen production is done inside GREET. In this analysis, we assumed that hydrogen would be produced from natural gas via steam methane reforming.

3. DATA SOURCES AND ASSUMPTIONS FOR GREET SIMULATIONS

3.1. Soybean Farming

3.1.1. Yield

Soybean yield (in bushels per acre or bu/acre) is a key factor in life-cycle analysis because it will affect energy use and fertilizer use per bushel of soybeans harvested. Soybeans were ranked the second-leading U.S. crop in terms of both harvested acreage (74.6 million acres) and revenue (19.7 billion U.S. dollars [USD]) in 2006 (USDA 2007a). Over the past several decades, both harvested acreage and soybean yield per harvested acre have experienced enormous growth, leading to total soybean production increases of 4% annually. Table 3-1 lists planted and harvested acreage and yield over the past five decades in the United States. Figure 3-1 shows the 3-year moving average of soybean yield in the United States. The soybean yield has been increasing at an annual rate of 1.2%, and this trend is expected to continue in the near future.

**Table 3-1. U.S. Historical Soybean Acreage and Yields
(USDA 2007a)**

Year	Acreage (10^6 acres)		Total Production (10^6 bu)	Yield (bu/acre)		3-Year Moving Average Yield (bu/acre)	
	Planted	Harvested		Planted Acres	Harvested Acres	Planted Acres	Harvested Acres
1950	15.0	13.8	299.2	19.9	21.7	19.5	21.8
1960	24.4	23.7	555.1	22.7	23.5	22.9	23.7
1970	43.1	42.2	1127.1	26.2	26.7	26.3	26.9
1980	69.9	67.8	1797.5	25.7	26.5	28.8	29.3
1990	57.8	56.5	1925.9	33.3	34.1	30.4	31.1
1991	59.2	58.0	1986.5	33.6	34.2	32.8	33.5
1992	59.2	58.2	2190.4	37.0	37.6	34.6	35.3
1993	60.1	57.3	1869.7	31.1	32.6	33.9	34.8
1994	61.6	60.8	2514.9	40.8	41.4	36.3	37.2
1995	62.5	61.5	2174.3	34.8	35.3	35.6	36.4
1996	64.2	63.3	2380.3	37.1	37.6	37.6	38.1
1997	70.0	69.1	2688.8	38.4	38.9	36.8	37.3
1998	72.0	70.4	2741.0	38.1	38.9	37.9	38.5
1999	73.7	72.4	2653.8	36.0	36.6	37.5	38.2
2000	74.3	72.4	2757.8	37.1	38.1	37.1	37.9
2001	74.1	73.0	2890.7	39.0	39.6	37.4	38.1
2002	74.0	72.5	2756.1	37.3	38.0	37.8	38.6
2003	73.4	72.5	2453.7	33.4	33.9	36.6	37.2
2004	75.2	74.0	3123.7	41.5	42.2	37.4	38.0
2005	72.0	71.3	3063.2	42.5	43.0	39.2	39.7
2006	75.5	74.6	3188.2	42.2	42.7	42.1	42.7

3.1.2. Energy Use

The USDA's Economic Research Service (ERS) survey data provides U.S. energy use values for soybean farming (on a per-acre basis) in 2002 (USDA 2007b); these values are listed in Table 3-2. On the basis of these energy use values and the average yields for soybeans, we estimated the energy use (by type) per bushel of soybeans harvested. We converted the values listed in Table 3-2 to Btu-based values by using the lower heating values (LHVs) of fuels in GREET: 128,450 Btu/gal for diesel; 116,090 Btu/gal for gasoline; 84,950 Btu/gal for liquefied petroleum gas (LPG); 3,412 Btu/kWh for electricity (energy loss for electricity generation is simulated separately in GREET); and 983 Btu/ft^3 for natural gas. The total energy use is estimated to be 22,084 Btu/bu: 64% diesel, 18% gasoline, 8% LPG, 7% natural gas, and 3% electricity. In comparison, Hill et al. (2006) reported 23,474 Btu/bu and 34,625 Btu/bu when custom-work-related diesel use and farm-related transportation and personal commuting energy use are taken into account. Pimentel and Patzek (2005) reported 20,447 Btu/bu of energy use for soybean production when labor, machinery, and fertilizer were taken into account. Table 3-3 provides a detailed comparison of the energy use for soybean farming across these references.

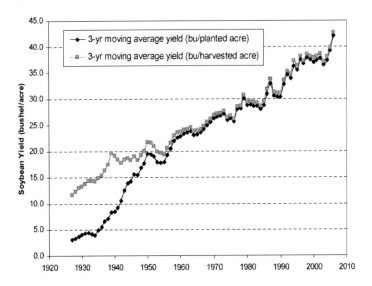

Figure 3-1. Three-Year Moving Average of Soybean Yield in the United States (USDA 2007a).

Table 3-2. Energy Use for Soybean Farming in the United States (USDA 2007b)

State	Diesel (gal/acre)	Gasoline (gal/acre)	LPG (gal/acre)	Electricity (kWh/acre)	Natural Gas (ft^3/acre)
Arkansas	9.9	1.3	L[a]	11.2	L
Illinois	2.5	0.9	0.0	L	0.0
Indiana	2.3	1.6	L	1.3	L
Iowa	3.4	1.1	0.0	0.0	0.0
Kansas	2.9	1.1	1.8	9.1	349.2
Kentucky	2.1	1.4	L	4.5	0.0
Louisiana	6.5	1.1	L	L	L
Maryland	2.9	2.1	L	0.8	0.0
Michigan	4.0	1.5	L	L	0.0
Minnesota	4.0	1.1	L	L	0.0
Mississippi	4.3	1.2	L	3.8	0.0
Missouri	4.3	1.4	L	L	0.0
Nebraska	12.9	1.3	4.4	39.4	586.4
North Carolina	2.4	1.5	L	0.6	0.0
North Dakota	3.2	1.4	L	0.8	0.0
Ohio	2.0	1.3	L	0.0	0.0
South Dakota	2.8	1.4	0.0	L	0.0
Tennessee	2.2	1.3	L	1.0	0.0
Virginia	1.9	1.2	L	L	0.0
Wisconsin	5.2	2.4	0.0	L	0.0
Average of all states	4.1	1.3	0.4	7.8	52.5
Energy use (Btu/bu)	14,221.8	3,934.1	1676.9	634.7	1619.9
Total energy use (Btu/bu)					22,087.4

[a] L = insufficient data for legal disclosure.

**Table 3-3. Comparison of Energy Use for Soybean Farming
Taken from Three Data Sources**

Parameter	Source		
	USDA 2007b	Hill et al. 2006	Pimentel and Patzek 2005
Year	2002	2002	Not available
Energy use (Btu/bu)	22,087	23,474/34,625[a]	20,447[b]
Percentage			
Diesel	64.4	61.7	57.7
Gasoline	17.8	17.2	35.2
LP gas	7.6	4.1	3.3
Electricity	2.9	11.0	3.8
Natural gas	7.3	6.1	0

[a] The 34,625 value includes diesel use of 6.6 L/ha for custom work and farm-related transportation and personal commuting energy use equal to those values associated with corn farming.
[b] Including energy input for labor, machinery, and fertilizer.

3.1.3. Fertilizer Use

We updated fertilizer use values for soybean farming in GREET by using the newly released USDA ERS data (USDA 2007c) (see Table 3-4). We used soybean yield per planted acre to calculate the fertilizer use per bushel of soybeans. Figure 3-2 shows the fertilizer use for soybean farming over the past 15 years. The amount of fertilizer used (nitrogen [N], phosphorous [P], and potassium [K], in grams) per bushel of soybeans did not change significantly. In fact, the usage patterns for each fertilizer type follow a similar time trend. For year 2010 (as our target year for this study), the following amounts were used: nitrogen at 61.2 g/bu, phosphorus at 186.1 g/bu, and potassium at 325.5 g/bu. The energy use and emissions for fertilizer manufacturing are simulated separately in GREET. On the basis of GREET simulations, the total energy use values per gram of fertilizer produced are 45.9 Btu/g N, 13.29 Btu/g P, and 8.42 Btu/g K.

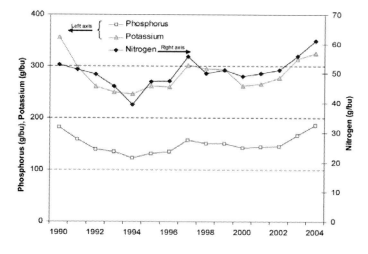

Figure 3-2. Fertilizer Use for Soybean Farming in the United States.

Table 3-4. Fertilizer Use for Soybean Farming (USDA 2007c)

Year	Percent Acreage Receiving Nitrogen Fertilizer	Nitrogen Application Rate (lb/received acre)	Percent Acreage Receiving Phosphorus Fertilizer	Phosphorus Application Rate (lb/received acre)	Percent Acreage Receiving Potassium Fertilizer	Potassium Application Rate (lb/received acre)
1988	16	22	26	48	31	79
1989	17	18	28	46	32	74
1990	17	24	24	47	29	81
1991	16	25	22	47	23	76
1992	15	22	22	47	25	75
1993	14	21	21	46	25	79
1994	13	25	20	47	25	82
1995	17	29	22	54	25	85
1996	15	24	25	49	27	85
1997	20	25	28	50	33	88
1998	17	23	24	48	27	81
1999	18	21	26	46	28	78
2000	18	24	24	48	27	76
2001	NA[a]	24	NA	49	NA	84
2002	20	21	26	49	29	89
2003	NA	NA	NA	NA	NA	NA
2004	21	28	26	69	23	121
2005	NA	NA	NA	NA	NA	NA

[a] NA = not available.

3.1.4. N2O Emissions

N2O, a potent GHG, is produced from nitrogen in the soil through nitrification and denitrification processes (direct N2O emissions). N2O can also be produced through volatilization of nitrate from the soil to the air and through leaching and runoff of nitrate into water streams (indirect N2O emissions).

Estimation of direct and indirect N2O emissions from crop farming requires two important parameters: (1) the amount of nitrogen applied to soil and (2) rates for converting nitrogen into N2O. The application of nitrogen fertilizer is the key to crop farming. For legume crops, such as soybeans, nitrogen fixation is another major nitrogen input. In 1996, IPCC considered nitrogen input to soil from biological nitrogen fixation by legume crops in estimating N2O emissions from soil. However, in 2006, IPCC elected not to consider this nitrogen input because of a lack of evidence of significant emissions from the nitrogen fixed by legumes.

Even without considering the nitrogen that results from the biological fixation process, two sources of nitrogen inputs to soil for crop farming remain: nitrogen from fertilizer application and nitrogen in the aboveground biomass left in the field after harvest and in the belowground biomass (i.e., roots). For crops such as corn, nitrogen in the aboveground and belowground biomass is from nitrogen fertilizers. For crops such as soybeans, nitrogen in the aboveground and belowground biomass is eventually from nitrogen fertilizers and the biological nitrogen fixation process. GREET 1.8 takes into account the nitrogen in nitrogen fertilizers and the nitrogen in aboveground and belowground biomass in estimating N2O emissions from crop farming.

For corn, IPCC (2006) estimates that aboveground biomass is 87% of corn yield (on a dry-matter basis). Aboveground biomass has a nitrogen content of 0.6%. Belowground biomass is about 22% of aboveground biomass, with a nitrogen content of 0.7%. The total amount of nitrogen in corn biomass that is left in corn fields per bushel of corn harvested is calculated as shown in Equation 3-1:

$$56 \text{ lb/bul} \times 85\% \text{ (dry matter content of corn)} \times (87\% \times 0.6\% + 87\% \times \quad (3\text{-}1)$$
$$22\% \times 0.7\%) = 0.312 \text{ lb N/bu} = 141.6 \text{ g/bu}$$

To estimate N2O emissions from corn farming, 141.6 g of N are added to nitrogen fertilizer inputs for corn farming (which are about 420 g of N per bushel).

For soybeans, IPCC (2006) states that aboveground biomass is about 91% of soybean yield (on a dry-matter basis). Aboveground biomass has a nitrogen content of 0.8%. Belowground biomass is about 19% of aboveground biomass, with a nitrogen content of 0.8%. The total amount of nitrogen in soybean biomass that is left in soybean fields per bushel of soybean harvested is calculated as shown in Equation 3-2:

$$60 \text{ lb/bu} \times 85\% \text{ (dry matter content of soybeans)} \times \quad (3\text{-}2)$$
$$(91\% \times 0.8\% + 91\% \times 19\% \times 0.8\%) = 0.442 \text{ lb N/bu} = 200.7 \text{ g/bu}$$

To estimate N2O emissions from soybean farming, 200.7 g of N are added to nitrogen fertilizer inputs for soybean farming (which are about 62 g of N per bushel). The rates for converting the nitrogen in soil and water streams to N2O emissions to the air are subject to great uncertainties (Wang et al. 2003; Crutzen et al. 2007). IPCC (2006) presents a conversion rate of 1% for direct N2O emissions from soil (compared with 1.25% in IPCC [1996]), with a range of 0.3–3%.

Indirect N2O emissions include those from volatilization of nitrate from the soil to the air and leaching and runoff of nitrate into water streams where N2O emissions occur. IPCC (2006) estimates a volatilization rate for soil nitrogen of 10%, with a range of 3–30%. The conversion rate of volatilized nitrogen to N in N2O emissions is 1%, with a range of 0.2–5%. The leaching and runoff rate of soil nitrogen is estimated to be 30%, with a range of 10–80%. The conversion rate of leached and runoff nitrogen to N in N2O emissions is 0.75%, with a range of 0.05–2.5%.

Thus, the conversion rate for direct and indirect N2O emissions is 1.325% (1% + 10% × 1% + 30% × 0.75%). This conversion rate was used in GREET 1.8. In contrast, Crutzen et al. (2007) estimated a conversion rate of 3–5% on the basis of the global N2O balance. While the top-down approach adopted in Crutzen et al. is a sound approach, especially for checking and verifying results against the bottom-up approach used by the IPCC and others, data for the top-down approach needs to be closely examined in order to generate reliable N2O conversion factors. In particular, Crutzen et al. adopted the global N2O emission balance from a 2001 study but adopted the nitrogen inputs from a separate 2004 study for deriving N2O conversion factors. Furthermore, Crutzen et al. did not get into agricultural subsystems (such as crop farming, animal waste management, and crop residual burning), which are required for generating N2O conversion rates for the nitrogen inputs into crop farming. Their allocation of aggregate N2O emissions (even after subtracting N2O emissions from industrial sources) to the aggregate agricultural system could result in overestimation of N2O conversion rates from nitrogen inputs into crop farming systems. Nonetheless, N2O conversion rates, which are subject to great uncertainties, need to be reconciled between the bottom-up and the top-down approach.

3.2. Soy Oil Extraction

At soybean processing plants, soybean seeds are crushed, soy oil is extracted from the crushed seeds, and crude soy oil is refined. Soybeans contain 18–20% oil by weight. To maximize soy oil production, organic solvents are used during oil extraction. The solvent extraction process is a widely used and well-established technology. The standard solvent extraction process uses n-hexane that is produced from petroleum. Most of the n-hexane used in oil extraction is recovered and recycled, with some inevitable loss. Table 3-5 presents the inputs and outputs from oil extraction plants. In calculating emissions and energy use, we assumed that steam is generated from natural gas. N-hexane is a straight-chain hydrocarbon. Commercial hexane is manufactured by distillation of straight-run gasoline produced from crude oil or natural gas liquids. In GREET, hexane is assumed to be produced from crude oil, and its upstream production energy use and emissions are adopted from energy use and emissions calculated for production of LPG from crude oil. Because hexane is volatile, the amount of hexane lost during soy oil extraction is assumed to be in the form of VOC emissions to the atmosphere. For more details, see Wang (1999).

Table 3-5. Inputs and Outputs of Soybean Oil Extraction Plants

Inputs and Outputs	GREET Value[a]	
Input		
Soybeans (lb)	5.7	
Steam (Btu)	2,900	(44.5%)
NG (Btu)	2,800	(43.0%)
Electricity (Btu)	614 (9.4%)	
N-hexane (Btu)	205 (3.1%)	
Total energy (Btu)	6,519	(100%)
Output		
Inputs and Outputs	GREET Value[a]	
Soy oil (lb)	1	
Soy meal (lb)	4.48	

[a] From previous GREET assumptions. We assumed in GREET that steam is produced from natural gas with an efficiency of 80%. The Btu value for steam is the natural gas Btu used to generate the needed steam. Values in parentheses are percentage shares of process fuels.

3.3. Production of Soybean-Derived Fuels

Figure 3-3 illustrates the fuel production processes for the four soybean-derived fuels.

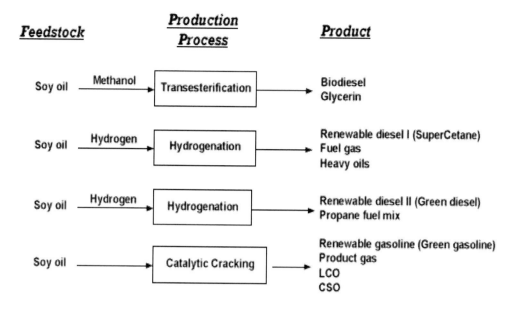

Figure 3-3. Fuel Production Processes for the Four Soybean-Derived Fuels.

3.3.1. Biodiesel

Biodiesel is produced through the so-called transesterification process, in which soy oil is combined with alcohol (ethanol or methanol) in the presence of a catalyst (sodium hydroxide [NaOH] in this case) to form ethyl or methyl ester, as illustrated in Figure 3-4. The transesterification process requires steam and electricity as energy inputs and produces both biodiesel and glycerin.

For this study, we updated GREET biodiesel production simulations on the basis of data in Haas et al. (2006). Table 3-6 presents the inputs and outputs of biodiesel plants per pound of biodiesel produced. To apply the values specified in Table 3-6 to GREET, we assumed that (1) steam is generated from natural gas with an energy conversion efficiency of 80% and (2) the energy embedded in the three chemical compounds is half oil and half natural gas.

$H_2C - OCOR'$			$ROCOR'$	$H_2C - OH$
$HC - OCOR''$	$+$	$3\,ROH$	$ROCOR''$	$+$	$HC - OH$
$H_2C - OCOR'''$			$ROCOR'''$	$H_2C - OH$
Triglyceride		Alcohol	Mixture of methyl esters	Glycerin

Figure 3-4. Transesterification of Soy Oil to Biodiesel.

Table 3-6. Inputs and Outputs of Biodiesel Plants (lb or Btu/lb biodiesel)

Inputs and Outputs	Haas et al. 2006	Sheehan 1998	GREET Value
Inputs			
Soy oil (lb)	1.001	1.050	1.001
Methanol (lb)	0.1001	0.0900	0.1001
Sodium hydroxide (lb)	0.0050	0.0023	0.0050
Sodium methoxide (lb)	0.0125	0.0244	0.0125
Hydrochloric acid (lb)	0.0071	0.0077	0.0071
NG (Btu)	888	789	888
Electricity (Btu)	46	45	46
Outputs			
Biodiesel (lb)	1	1	1
Glycerin (lb)	0.116	0.213	0.213

3.3.2. Renewable Diesel I

The production of renewable diesel I comprises a series of reactions, including those involved in hydrocracking (breaking apart of large triglyceride molecules), hydrotreating (removal of oxygen), and hydrogenation (saturation of double bonds). Besides soy oil, hydrogen is needed as input. Some steam is also needed; ASPEN simulations conducted by NREL assumed that the required steam would be generated with the fuel gas and/or heavy oils that are co-produced from the plant. The output of this process is high-cetane diesel (with fuel gas and heavy oils as co-products). Table 3-7 lists the inputs and outputs of renewable diesel I plants. Note that the output values for fuel gas and heavy oils are net amounts (i.e., after steam generation for internal use). In GREET, hydrogen used in renewable diesel plants is assumed to be produced from natural gas via steam methane reforming (SMR).

**Table 3-7. Inputs and Outputs of Renewable
Diesel I Plants (lb or Btu per lb of renewable diesel I)**

Inputs and Outputs	ASPEN Simulation Results as GREET Input
Inputs	
Soy oil (lb)	1.510
Hydrogen (lb)	0.030
Electricity (Btu)	134.4
Outputs	
Renewable diesel I (lb)	1
Fuel gas (Btu)	7083.7
Heavy oils (Btu)	3608.0

3.3.3. Renewable Diesel II

For the production of renewable diesel II, soy oil is combined with hydrogen in a catalytic reactor and then converted by a hydrogenation reaction to a high-cetane renewable diesel. This process requires electricity and thermal energy as inputs; the outputs are renewable diesel and a small amount of propane fuel mix. We assumed that thermal energy is generated from natural gas with an energy conversion efficiency of 80% and that hydrogen is produced from natural gas via SMR. Table 3-8 presents the inputs and outputs of renewable diesel plants per pound of renewable diesel II produced.

**Table 3-8. Inputs and Outputs of Renewable Diesel II
Plants (lb or Btu per lb of renewable diesel II)**

Inputs and Outputs	ASPEN Simulation Results as GREET Input
Inputs	
Soy oil (lb)	1.174
Hydrogen (lb)	0.032
Natural gas (Btu)	84.05
Electricity (Btu)	93.83
Outputs	
Renewable diesel II (lb)	1
Propane fuel mix (Btu)	1095.5

3.3.4. Renewable Gasoline

The production of renewable gasoline takes place in an FCC unit. This process requires electricity and steam. The steam is assumed to be generated by combusting the by-product and product gas mix that results from the cracking process. The process also generates extra steam for export. The outputs are renewable diesel, product gas, LCO, and CSO. Table 3-9 presents the inputs and outputs from renewable gasoline plants per lb of renewable gasoline produced.

**Table 3-9. Inputs and Outputs of Renewable
Gasoline Plants (lb or Btu per lb of renewable gasoline)**

Inputs and Outputs	Aspen Simulation Results as GREET Input
Inputs	
Soy oil (lb)	2.231
Electricity (Btu)	185.6
Outputs	
Renewable gasoline (lb)	1
Product gas (Btu)	6313.5
LCO (Btu)	4737.4
CSO (Btu)	5460.3

3.3.5. Comparison of the Four Soybean-Derived Fuels

On the basis of the analysis and assumptions outlined in Sections 3.3.1 through 3.3.4, Table 3-10 summarizes the energy use and amounts of product and co-product that can be produced from 1 ton of soybeans. According to Table 3-10, the transesterification process can generate a much larger amount of diesel product and co-products from 1 ton of soybeans than the other processes; however, it requires a lot more energy and chemical inputs than do the other processes. The hydrogenation process (used to produce renewable diesel II) has the best yield (in terms of energy content from 1 ton of soybeans) of the three new fuels, while it generate less energy co-product than the other processes. Because all of the processes produce other products (besides the target fuel), the energy value or market value of the co-products of these processes is an important factor in evaluating the energy and emission benefits of each soybean-based fuel. The co-product issue is discussed in Section 4. The production processes for the two renewable diesel options require hydrogen. Because hydrogen production is energy intensive, so determining which process is more energy intensive simply on the basis of inputs and outputs would not lead to a proper conclusion. The fuel cycles of hydrogen and other types of energy inputs must be taken into consideration, emphasizing the importance of a complete life-cycle analysis like the one conducted for this study.

Table 3-10. Energy Use and Amount of Fuel Product and Co-Products from One Ton of Soybeans

Inputs and Outputs	Fuel			
	Biodiesel	Renewable Diesel I	Renewable Diesel II	Renewable Gasoline
Outputs				
Product lb	351	232	299	157
mmBtu	5.66	4.36	5.66	2.94
Co-products				
Soy meal (lb)	1572	1572	1572	1572
Glycerin (lb)	75			
Energy co-product (mmBtu)		2.48	0.33	2.60
Inputs				
Natural gas (mmBtu)				
Soy oil extraction	1.80		1.80	
Fuel production	0.31		0.03	
Electricity (mmBtu)				
Soy oil extraction	0.194	0.194	0.194	0.194
Fuel production	0.016	0.031	0.028	0.029
Other inputs				
Methanol (mmBtu)	0.303			
Hydrogen (mmBtu)		0.36	0.49	

3.4. Fuel Properties

Table 3-11 presents the properties of the soybean-based fuels examined in this study. Compared with conventional diesel and biodiesel, renewable diesel fuels have much higher cetane numbers and lower density. Cetane number is one measure of the quality of a diesel fuel — a high number is a valuable feature for renewable diesel as a diesel blending component and a cetane enhancer.

3.5. Fuel Use in Vehicles

For our life-cycle analysis, we assumed that soybean-derived diesel fuels are used in 100% pure form in compression-ignition, direct-injection (CIDI) engine vehicles, and renewable gasoline is used in 100% pure form in spark-ignition (SI) engine vehicles. Since there were no testing data, we assumed that the fuel economy and CH4 and N2O emissions for CIDI vehicles are the same for all three diesel types. Likewise, we assumed that the fuel economy and CH4 and N2O emissions for SI vehicles are the same for the two gasoline types.

Table 3-11. Properties of the Four Soybean-Based Fuels

Fuel	Lower Heating Value (Btu/gal)	Density (lb/gal)	Carbon Content (%)[e]	Oxygen Content (%)	Cetane Value
Petroleum gasoline[a]	113,602	6.23	84.0	NA[f]	NA
Petroleum diesel[a]	129,488	7.06	87.1	0.0	40
Biodiesel[a]	119,550	7.40	77.6	11.0	50–65
Renewable diesel I[b]	117,059	6.24	87.1	0.0	100
Renewable diesel II[c]	122,887	6.49	87.1	0.0	70–90
Renewable gasoline[d]	115,983	6.21	84.0	NA	NA

[a] From the GREET model.
[b] From (S&T)[2] Consultants Inc. (2004).
[c] From Kalnes et al. (2007).
[d] From UOP (2005).
[e] Because of a lack of data, the carbon content of renewable diesel fuels is assumed to be the same as that for petroleum-based diesel; the carbon content of renewable gasoline is assumed to be the same as that of petroleum-based gasoline.
F NA = not applicable.

4. CO-PRODUCT CREDITS FOR BIOFUELS

4.1. Methods for Addressing Co-Product Credits

The objective of calculating the credit allotted for co-products in life-cycle analysis is to fairly address the energy and emission burdens of the primary product, especially when the co-products have value in the marketplace. Two methods that are commonly used are the displacement method and the allocation method.

With the displacement method, a conventional product is assumed to be displaced by a new product. The life-cycle energy that would have been used and the emissions that would have been generated during production of the displaced product are counted as credits for the new product that is co-produced from the fuel pathway under evaluation. These credits are subtracted from the total energy use and emissions associated with the fuel pathway under evaluation. The difficulties with the displacement method involve accurately determining the displaced products and identifying the approach to obtain their life-cycle energy use and emissions. Also, if the amounts of co-products are relatively large compared with the amount of primary product from a given process (as is the case for renewable diesel I and renewable gasoline, see Table 3-10), the displacement method results — which are WTW analysis results that are mathematically normalized to production of a unit of the primary product — can generate distorted results for the primary product.

The allocation method allocates the feedstock use, energy use, and emissions between the primary product and co-products on the basis of mass, energy content, or economic revenue. This method is easier to implement in life-cycle analyses than the displacement method. However, it could result in inaccurate results if the values of product and co-products cannot be simply measured on a single basis (such as mass or energy content).

In this study, various co-products are produced during the production of soybean-based fuels, including protein products such as soy meal; solvents such as glycerin; and energy products such as propane fuel mix and heavy oils (see Table 3-10), which makes addressing their credit very difficult. If the displacement method is used, it is time-consuming to identify a displaced product for each of the co-products and obtain the life-cycle energy use and emissions of the identified products. Besides, the co-products almost have Btu values equivalent to those of their primary products (e.g., renewable diesel I and renewable gasoline), which makes the displacement method not a preferable approach. On the other hand, because these co-products have different values (for instance, the primary products and most of the co-products have Btu values and can be treated as energy products; some of the co-products, however — such as soy meal and glycerin — have nonenergy values), the Btu-based allocation method would not be able to fairly treat the co-products that have low energy contents but are valuable in other ways. The market value-based allocation method is subject to the variation in price of the co-products.

Table 4-1. Approaches to Address Co-Products of Soybean-Based Fuels

Fuel Product	Process	Approach 1 (Displacement)	Approach 2 (Energy-Value-Based Allocation)	Approach 3 (Market Value-Based Allocation)	Approach 4 (Hybrid)
Biodiesel	Soy oil extraction	Displacement	Allocation	Allocation	Displacement
production	Transesterification	Displacement	Allocation	Allocation	Displacement
Renewable diesel I	Soy oil extraction	Displacement	Allocation	Allocation	Displacement
production	Hydrogenation	Displacement	Allocation	Allocation	Allocation
Renewable diesel	Soy oil extraction	Displacement	Allocation	Allocation	Displacement
II production	Hydrogenation	Displacement	Allocation	Allocation	Allocation
Renewable gasoline	Soy oil extraction	Displacement	Allocation	Allocation	Displacement
production	Catalytic cracking	Displacement	Allocation	Allocation	Allocation

On the basis of these considerations, four approaches were employed to address the co-product issues: (1) the displacement approach, (2) an energy-based allocation method, (3) an allocation method based on the market values of the primary products and co-products, and (4) a hybrid approach that employs both the displacement and the allocation methods, in which the displacement method is used for soy meal and glycerin, and the allocation method is used for other energy co-products. For biodiesel, the hybrid approach is the same as the displacement approach. Table 4-1 summarizes the four approaches.

4.2. Displacement Approach

The first step in using the displacement method is to determine an equivalent product replaced by each co-product. Soy meal, which is primarily used as a livestock feed in the United States, is assumed in this study to replace soybeans. Soybean-based glycerin is assumed to replace petroleum-based glycerin. Other energy co-products are assumed to replace similar energy forms on the basis of their energy value; for example, fuel gas is assumed to replace equivalent-Btu natural gas for industrial use, heavy oil is assumed to replace equivalent-Btu residual oil. Table 4-2 lists the products that are to be displaced by the co-products from soybean-based fuel production.

Table 4-2. Products to Be Displaced by Co-Products

Product	Product to Be Displaced
Soy meal	Soybeans
Glycerin	Petroleum-based glycerin
Fuel gas	Natural gas
Heavy oil	Residual oil
Propane fuel mix	LPG
Product gas	Natural gas
LCO	Diesel fuel
CSO	Residual oil

The energy use and emissions resulting from production of one million Btu of natural gas, residual oil, LPG, and diesel fuel are already simulated in GREET and can be readily used. Also, GREET has addressed life-cycle energy use and emissions for obtaining soybeans, including soybean farming and fertilizer manufacturing, and these results are also readily used.

However, the displacement ratio between soy meal and soybeans for the purpose of feeding animals is yet to be determined in our study. Moreover, life-cycle analysis for petroleum-based glycerin is not included in GREET and thus needs further examination in this study.

4.2.1. Soy Meal

The displacement ratio of soy meal to soybeans is determined by protein content. Literature reports a protein content of 44–50% in soybean meal and 35–40% in soybeans (Ahmed et al. 1994; Maier et al. 1998; Britzman 2000). In this study, we assumed that soy meal contains 48% protein and soybeans contain 40%. On the basis of that assumption, we estimated that 1 lb of soy meal can replace 1.2 lb of soybeans.

4.2.2. Glycerin

Glycerin produced from petrochemical sources is called synthetic glycerin; natural glycerin is produced from plant oils and animal fats. Petroleum-based glycerin uses propylene, chlorine, and sodium hydroxide as raw materials. The theoretical raw material input to produce 1 lb of glycerin can be calculated according to the mass balance of the chemical reactions. In practice, there are some differences between theoretical mass balance and actual plant mass balance. Table 4-3 shows the amount of raw material needed to produce 1 lb of synthetic glycerin.

Table 4-3 Raw Material Input for One Pound
of Synthetic Glycerin (lb/lb glycerin)

	Theoretical Input[a]	Industry Input[b]
Propylene	0.46	0.62
Chlorine	1.54	2.00
Sodium compounds	0.87	0.90

[a] Based on Chemical Economics Handbook (Greiner et al. 2005; Malveda et al. 2005).
[b] From Ahmed et al. (1994).

Production of synthetic glycerin requires little energy, so this energy is not addressed in our analysis. The energy use and emissions embedded in the raw material are the key issues in determining the life-cycle energy use and emissions of synthetic glycerin.

In this study, the production data for propylene, chlorine, and sodium hydroxide were taken from the Eco-Profile life-cycle inventory (Association of European Plastic Industry 2005). The Eco-Profile reports average industry data in detail for various petrochemical processes, including the amount of petroleum and natural gas used as feedstocks to produce each type of chemical, and the amount of petroleum, natural gas, electricity, and other fuels used as process fuels. We use the GREET model to generate the upstream energy use and emissions for the fuel (e.g., petroleum, natural gas, and electricity) used in producing propylene, chlorine, and sodium hydroxide. Table 4-4 compares the total energy embedded in raw material per pound of glycerin between our study and the study conducted by Ahmed et al. Some European studies report 30,000 to 90,000 Btu of total or fossil energy (Scharmer and Gosse 1996; Malça and Freire 2006).

Table 4-4. Total Btu in Raw Material per Pound of Glycerin

Study	Propylene	Chlorine	Sodium Hydroxide	Total
Our study	9,373	12,267	10,128	39,460
Ahmed et al. (1994)	8,577	5,319	11,275	21,296

4.3. Allocation Approach

Two different allocation approaches are applied in this study: energy-value-based and market- value-based. Generally, the allocation method is easier to implement than the displacement method in terms of data requirements. With the energy-value-based allocation method, the energy contents of the primary product and co-products are used to split the burden of energy input, feedstock input, and pollutant emissions. With the market-value-based allocation method, the market value of the products becomes the determining factor in splitting the burden.

4.3.1. Allocation at the System Level and Subsystem Level

The process of producing soybean-based fuels from soybeans involves two stages: soy oil extraction and fuel production. Both stages generate co-products, resulting in two different ways of allocating co-product credit: system level and subsystem level.

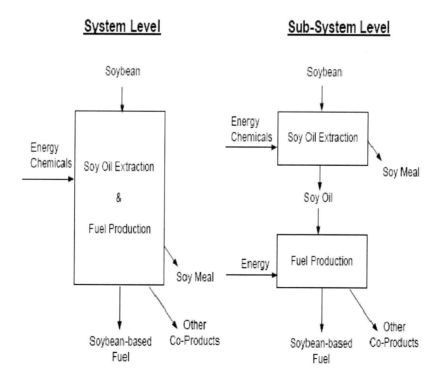

Figure 4-1. Two System Levels of Soybean-Based Fuel Production in the Allocation Approach.

As Figure 4-1 shows, system-level allocation takes soy oil extraction and fuel production processes as a whole system, with soybeans and the required energy and chemicals as inputs and fuel, soy meal, and other co-products as outputs. With the whole system level, the effect of soy oil is eliminated. Subsystem-level allocation includes two subsystems. In the first, soybeans are the inputs, and soy oil and soy meal are the outputs; in the second, soy oil is the input.

The displacement method will give the same final results no matter which system level is considered, but the allocation method will not. Because the allocation ratio is determined by the energy value or market value of the primary product and co-products, the variation in market value of soy oil could obviously affect the allocation results of the first subsystem level but not affect the result of the second subsystem level, which means that it could affect the final results. However, in the soybean-to-biodiesel/renewable fuels case, soy oil is only a transitional product, which is produced and then consumed, so there is no reason that its market value or other value could affect the final results. On the basis of this consideration, we selected the whole-system level for the allocation approach.

4.3.2. Energy Value and Market Value

As mentioned, the energy value and market value of the primary product and co-products are the major determining factors for splitting energy and emissions among these products by using the allocation method. The energy value of soy meal was obtained from the Soybean Meal Info Center (http://www.soymeal.org). Note that soy meal is an animal food rather than a fuel, so its energy value is measured as the energy released when it is digested. The energy content of renewable fuels and their co-products were obtained on the basis of ASPEN simulation results (see Section 2.4).

Unlike the energy content value — which is stable and will not change — the market value of products could vary over time and by region. For soy meal, we used the average growth rate of the state-average market price during the last decade (1997–2007) to project market prices in 2010 (Ash and Dohlman 2007).

The glycerin market is heavily oversupplied worldwide (Malveda et al. 2005), so the price for glycerin is not expected to rise in the near future; in fact, extensive biodiesel production could even lower glycerin's market price. We assumed a price of $0.15/lb for glycerin, as provided in the Haas et al. (2006) study.

Because of the high cost of feedstock, the production cost of biodiesel is higher than that of conventional petroleum diesel. A wealth of research has been conducted to examine the cost for producing biodiesel at different industry scales (Haas et al. 2006; Bender 1999). These researchers estimate a production cost of $2.00–$2.30 per gallon of pure biodiesel, taking credits for soy meal and glycerin into consideration. The cost of biodiesel could vary significantly as a result of soybean and soy meal price variations. The United States has recently begun providing incentives to make biodiesel production costs competitive with those of petroleum-based diesel. Also, as biodiesel use increases and the infrastructure is established, the price of biodiesel could decrease. In this study, we used the biodiesel price before incentives.

For renewable diesel and gasoline fuels that are not yet on the market, we assumed the same market value as that of biodiesel fuel (on a per-million-Btu basis). Because the co-products of renewable diesel and gasoline production all have energy value and can be used in industry, we assumed the same prices per million Btu as their corresponding fuel (natural gas, residual oil, diesel, and LPG), determined as in Table 4-2. DOE's Energy Information Administration (EIA) *Annual Energy Outlook 2007* (EIA 2007a) projected the prices of natural gas, residual oil, diesel, and LPG in the industrial sector in 2010; these projected prices are used in our study.

Table 4-5 summarizes the energy content and market value of all products involved in this study. Note that prices in Table 4-5 are normalized to 2005 U.S. dollars (2005$) on the basis of an implicit U.S. price deflator from 1997 to 2006, as reported in the EIA *Annual Energy Review* (EIA 2007b).

Table 4-5. Energy Content and Market Value of Primary Products and Co-Products

Product or Co-Product	Energy Content (Btu/lb)	Market Value ($ 2005/lb)
Biodiesel	16,149	0.490
Renewable diesel I	18,746	0.569
Renewable diesel II	18,925	0.574
Renewable gasoline	18,679	0.567
Soy meal	4,246	0.274
Glycerin	7,979	0.150
Fuel gas	27,999	0.174
Heavy oils	20,617	0.195
Propane fuel mix	18,568	0.301
Product gas	18,316	0.114
LCO	19,305	0.248
CSO	18,738	0.177

4.3.3. Allocation Ratios

Table 4-6 presents the allocation ratios for the energy and emission burdens between primary products and co-products for the four soybean pathways. As indicated in Table 4-6, the allocation ratios of primary products based on energy value are a little lower than those based on market value.

4.4. Hybrid Approach

There are some shortcomings to both the displacement and allocation approaches. First, the production processes for renewable diesel I and renewable gasoline generate a large amount of co-products, resulting in overestimation of credits for those products if the displacement method is used. In fact, using this method can even result in negative energy input and emissions. On the other hand, in the energy-based allocation method, soy meal and glycerin have values not because they have energy content but for their other applications. Soy meal, particularly, has low energy value but high protein content and is thus valuable in the animal feed market; if soy meal is treated as fuel (like other energy co-products), its credit could be greatly underestimated. The market-value-based allocation method is subject to variations in the product prices, which may lead to numerous uncertainties.

Table 4-6. Allocation Ratios of Total Energy and Emission Burdens between Primary Products and Co-Products from Using the Allocation Approach (shown as %)

Product or Co-Product	Biodiesel	Renewable Diesel I	Renewable Diesel II	Renewable Gasoline
Energy-value-based allocation				
Primary fuel (biodiesel, renewable fuels)	42.9	32.2	44.7	24.1
Co-products (soy meal, glycerin, and others)	57.1	67.8	55.3	75.9
Market-value-based allocation				
Primary fuel (biodiesel, renewable fuels)	45.7	39.4	47.4	29.9
Co-products (soy meal, glycerin, and others)	54.3	60.6	52.6	70.1

To overcome these shortcomings, we introduced a hybrid approach, in which the displacement method is used for soy meal and glycerin, and the energy-based allocation method is used for other energy co-products. For biodiesel, the hybrid approach is the same as the displacement approach. Unlike the allocation approach, which considers the production processes from soybean to fuel as a whole system, the hybrid approach separates the production system into two subsystems because each subsystem is addressed by using different allocation methods. Table 4-7 presents the allocation ratio between primary products and co-products of the second subsystem that results from using the hybrid approach.

Table 4-7. Allocation Ratios of Total Energy and Emission Burdens between Primary Products and Co-Products of the Second Subsystem from Using the Hybrid Approach (%)

Parameter	Renewable Diesel I	Renewable Diesel II	Renewable Gasoline
Primary fuel (renewable fuels)	63.7	94.5	53.1
Co-products (heavy oil, etc)	36.3	5.5	46.9

5. LIFE-CYCLE ENERGY AND GHG EMISSION RESULTS FOR SOYBEAN-DERIVED FUELS

On the basis of the data and key assumptions presented in Section 3 and Section 4, we used GREET to conduct life-cycle simulations of energy use and GHG emissions for the six pathways examined in this study. GHG emissions are the sum of emissions of three gases — CO_2, CH_4, and N_2O — weighted by their global warming potentials. According to IPCC, the global warming potentials of CO_2, CH_4, and N_2O are 1, 25, and 298, respectively.

Figure 5-1 shows the GREET WTW modeling boundary. Results of a WTW analysis are separated into two stages: well-to-pump (WTP) and pump-to-wheels (PTW). Well-to-pump stages start with fuel feedstock recovery and end with fuels available at refueling stations. Pumpto-wheels stages cover vehicle operation activities. For example, for gasoline, the simulated stages include crude recovery; transportation of crude oil from oil fields to central storage terminals; crude oil storage at terminals; crude oil transportation from terminals to petroleum refineries; crude oil storage at refineries; crude refining to gasoline; transportation, storage, and distribution of gasoline; and combustion of gasoline in vehicles.

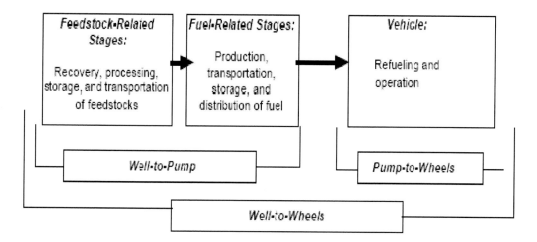

Figure 5-1. GREET Well-to-Pump and Pump-to-Wheels Stages.

In the following sections, petroleum-based RFG is the baseline for soybean-based renewable gasoline, and petroleum-based LSD is the baseline for soybean-based biodiesel and renewable diesel fuels.

5.1. Total Energy Use

Figure 5-2 presents WTW total energy use for 1 million Btu of fuel produced and used. Total energy use comprises all energy sources, including fossil energy and renewable energy (excluding energy embedded in soybeans, which is eventually from solar energy).

Figure 5-2 shows that different allocation approaches provide different results. The displacement approach gives the lowest total energy use among the four allocation approaches except in the case of renewable diesel II, whose production process generates a much smaller amount of co-product than the others. With the displacement approach, soybean-based fuels offer 6–25% lower total energy use than petroleum diesel or gasoline per million Btu, again except in the case of renewable diesel II, for which WTW total energy increases by 29% relative to LSD.

The two allocation approaches — energy-based allocation and market-based allocation — show good agreement with each other, with very similar results (1–4% difference). With the two allocation approaches, soybean-based fuels have 13–18% higher total energy use than petroleum diesel or gasoline.

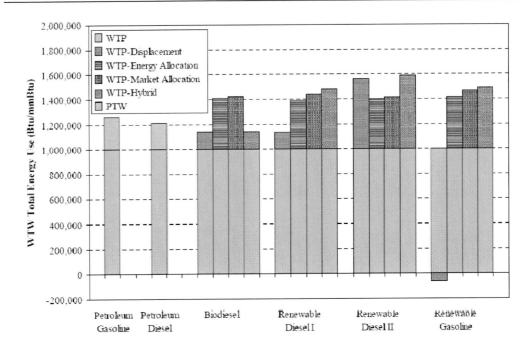

Figure 5-2. Well-to-Wheels Total Energy Use of Six Fuel Types.

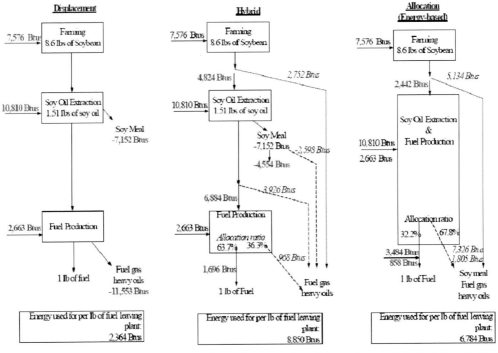

Note: Red indicates energy values allocated to primary product; blue values and dashed lines indicate energy values allocated to co-products.

Figure 5-3. Comparison of Total Energy Use among Three Allocation Approaches for Renewable Diesel I.

The hybrid approach gives the highest total energy use results for the renewable diesel and gasoline, 19–31% higher than their conventional counterparts. Biodiesel is an exception because the hybrid approach is exactly the same as the displacement approach for biodiesel. It is interesting that the hybrid approach provides higher energy use results than the displacement and allocation approaches, because the hybrid approach is derived from the integration of the both of the latter methods. To explore the reason, Figure 5-3 compares the allocation of energy use per pound of fuel leaving the plant for the three allocation approaches, taking renewable diesel I as an example. Note that the energy use in Figure 5-3 includes farming, transportation of feedstock, and production in the plant only, not over the whole life cycle. The higher energy use of the hybrid approach compared with the displacement approach is attributable to two factors. First, the farming and production energy use allocated to the final co-products (fuel gas and heavy oil) is much lower than their displacement credit (2,752 + 3,926 + 968 for the hybrid method versus 11,533 for the displacement method). Second, part of the credit for soy meal (-2,598) is allocated to the co-product (fuel gas and heavy oil), while all soy meal credit belongs to the primary product with the displacement approach. The reason that energy use is higher for the hybrid approach than the allocation approach is because the allocation approach allocates more energy to the co-products (5,134 + 7,326 + 1,805 for the allocation method versus 2,752 + 3,926 + 968 for the hybrid method) because the allocation ratio for co-products is much higher with the soy meal included (67.8% allocation versus 36.3% hybrid), and the difference between them (6,619) is larger than the soy meal credit earned in the hybrid approach (-4,554).

Renewable diesel II has fewer co-products; thus, its co-products and the method used to address them have a smaller effect on the results, which is apparent from the very similar energy use results among the four allocation approaches for this fuel.

5.2. Fossil Energy Use

Figure 5-4 presents the WTW fossil energy use of the six fuel options on the basis of 1 million Btu of fuel produced and used. Fossil energy use includes petroleum, natural gas, and coal.

Figure 5-4 reveals that all soybean-derived fuels offer significant reductions (52–107%) in fossil energy use. These reductions result from the fact that soybeans, as the feedstock for the four renewable fuel options, are a nonfossil feedstock. Soybean-based fuels, even with a certain amount of fossil energy input when they are used as process fuels during soybean farming and fuel production processes, can still achieve substantial reductions in fossil energy use.

Like the results for total energy use, the results for fossil energy use vary on the basis of the allocation method applied. With the displacement method, renewable gasoline can reduce WTW fossil energy use by 107% compared with petroleum gasoline. This large reduction in fossil energy use results from the large amount of co-products produced with renewable gasoline; these products were assumed to displace fossil energy (product gas to replace natural gas, LCO to replace diesel fuel, and CSO to replace residual oil), which helps renewable gasoline earn a large credit in fossil energy saving. Biodiesel, renewable diesel I, and renewable diesel II can achieve WTW fossil energy reductions of 84%, 90%, and 55%, respectively.

With the allocation approach, the reduction ratios are around 63–71%. The hybrid approach shows a 52–61% reduction in fossil energy use for soybean-based renewable fuels compared with conventional fuels.

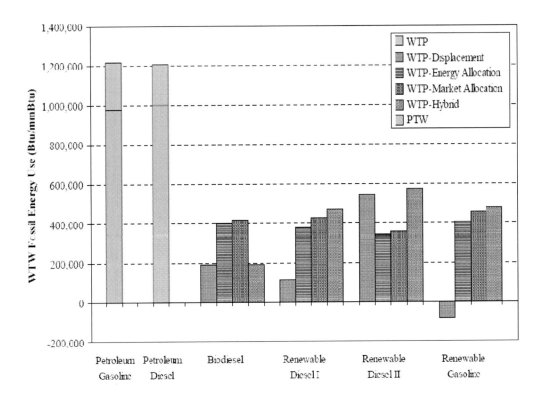

Figure 5-4. Well-to-Wheels Fossil Energy Use of the Six Fuel Types.

5.3. Petroleum Use

Figure 5-5 presents the WTW petroleum energy use for the six fuel options. Soybean-derived fuels offer significant oil savings. Petroleum energy used in the soybean-based fuel cycle is entirely from the WTP stage, primarily from diesel use for farming equipment and for the trucks and locomotives needed to transport feedstock and fuel. For soybean-based fuels, PTW fuel use is zero.

All of the four soybean-derived fuels can save more than 85% of petroleum use. With the displacement approach, for each million Btu of fuel produced and used, renewable gasoline reduces petroleum use by 148% compared with petroleum gasoline, and soybean-based diesel fuels reduce petroleum use by 99–106% relative to petroleum diesel. Like fossil energy use, the petroleum use associated with renewable gasoline is low because its production process generates large quantities of co-products (product gas, LCO, and CSO) in terms of Btu, and the co-products (LCO and CSO) are assumed to replace petroleum fuels (diesel and residual oil), providing large petroleum savings credits.

With the allocation approach, petroleum use among the four soybean-based fuels is very similar; use by all is about 88–92% lower than that of conventional petroleum fuels.

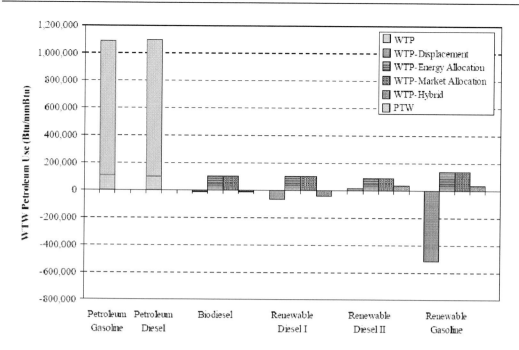

Figure 5-5. Well-to-Wheels Petroleum Energy Use of the Six Fuel Types.

With the hybrid approach, soybean-based fuels reduce WTW petroleum use by 97–104% relative to petroleum fuels. Unlike total energy use and fossil energy use results, WTW petroleum use for the hybrid approach is lower than that for the allocation approach for the three renewable fuels. This is because the production process for renewable fuels uses very little petroleum, so petroleum use allocated to the co-products is very small. On the other hand, farming of soybeans, assigned to be displaced by soy meal, consumes large amounts of diesel and gasoline, and makes the hybrid approach result in lower petroleum use because of the petroleum credit from soy meal.

5.4. GHG Emissions

Figure 5-6 presents WTW CO_2-equivalent grams of GHGs (including CO_2, CH_4, and N_2O) for the six fuel pathways studied. To clearly show the GHG reduction benefit of different soybean-based fuels, Figure 5-7 presents the changes in GHG emissions of the soybean-based fuels relative to their petroleum counterparts.

The emission results for the two renewable diesel fuels depend on the allocation approach used. Of the four allocation approaches, the displacement approach offers the best GHG reduction benefit, except for renewable diesel II. When this approach is used, all four soybean-based fuels can achieve a modest to significant reduction in WTW GHG emissions (64–174%) compared with petroleum-based fuels. The reason that renewable diesel I and renewable gasoline can achieve a much larger GHG emission reduction (-130% and –174%) is because they have a significant amount of co-products (fuel gas and heavy oil; product gas, LCO, and CSO) and because the production and combustion of the replaced fuels (natural gas, diesel fuel, and residual oil) could release lots of GHGs.

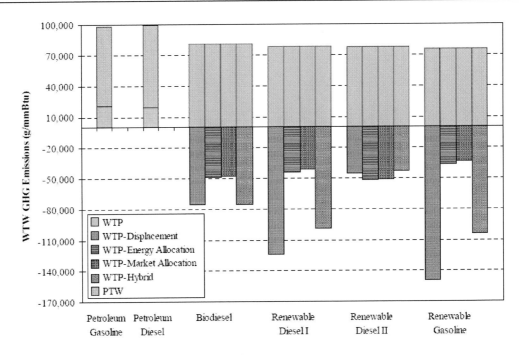

Figure 5-6. Well-to-Wheels GHG Emissions of the Six Fuel Types.

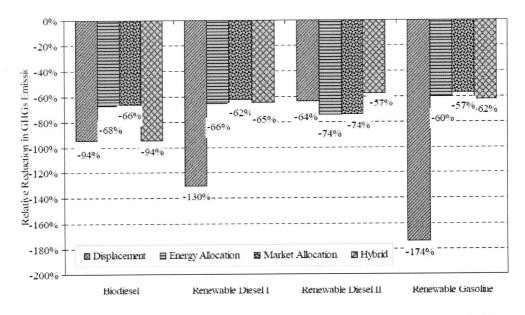

Figure 5-7. Well-to-Wheels GHG Emission Reductions for Soybean-Derived Fuels Compared with Petroleum Gasoline or Diesel.

With the allocation approach, soybean-based fuels achieve a modest reduction in GHG emissions (57–74%). The results from using the hybrid approach are similar to the results obtained from using the allocation approach.

These results are based on 1 million Btu of fuel produced and used. While we do not expect significant engine efficiency differences between the two gasoline types in SI engines and among the four diesel types in CIDI engines, it is well known that CIDI engines are more efficient than SI engines. Fuel consumption in CIDI engines could be 15–20% less than that of SI engines per distance traveled. To compare WTW results on a per-mile basis among the six options, researchers could reduce energy use and GHG emissions for the four diesel fuel options as presented in Figure 5-6.

CONCLUSIONS

We assessed the life-cycle energy and GHG emission impacts of soybean-derived biodiesel and soybean-derived renewable diesel and gasoline fuels by expanding, updating, and using the GREET model. Soybean-derived renewable diesel is produced from hydrogenation of soy oil, and renewable gasoline is produced from catalytic cracking of soy oil.

The method applied to determine energy and emission credits for co-products is a key issue in life-cycle analysis. The production processes of the four soybean-based fuels generate various kinds of co-products, which could lead to very different results depending on the method that is used to address the co-products. We used four different allocation approaches in this study: displacement, energy-based allocation, market-value-based allocation, and a hybrid approach (integrating the displacement and allocation methods). The four allocation approaches generate considerably different results.

For WTW total energy use, the displacement approach gives the lowest total energy use for the four bio-based fuels — showing a 6–25% reduction in total energy use for the biofuels (except for renewable diesel II) compared with petroleum fuels. The two allocation approaches show good agreement with each other, providing very similar results. The hybrid approach gives the highest total energy use results. Both the allocation and hybrid approaches show a 13–31% increase in total energy use compared with petroleum fuels.

All soybean-derived fuels achieve a significant reduction (52–107%) in fossil energy use. The displacement approach offers the best benefit in fossil energy use, with a reduction of 55–107%. With the allocation approach, the reduction ratios are around 63–71%. The hybrid approach shows a 52–61% reduction in fossil energy use for soybean-based renewable fuels compared with conventional fuels.

All four of the soybean-derived fuels can save more than 85% of petroleum use. With the displacement approach, renewable gasoline reduces petroleum use by 148% compared with petroleum gasoline because its production process generates a large amount of energy co-products. Soybean-based diesel fuels reduce petroleum use by 99–106% relative to petroleum diesel. With the allocation approach, the use of petroleum by the four soybean-based fuels is about 88–92% lower than its use by conventional petroleum fuels. With the hybrid approach, soybean-based fuels reduce WTW petroleum use by 97–104% relative to petroleum fuels.

With the displacement approach, all four soybean-based fuels can achieve a modest to significant reduction in WTW GHG emissions (64–174%) compared with petroleum-based fuels. While with the allocation approach, soybean-based fuels achieve a modest reduction in GHG emissions (57–74%).

REFERENCES

Ahmed, I., J. Decker, and D. Morris, 1994, *How Much Energy Does It Take to Make a Gallon of Soydiesel?* prepared for the National SoyDiesel Development Board.

Ash, M., and Dohlman, E., 2007, *Oil Crops Outlook: Prices Strengthen as Outlook for U.S. Soybean Supply Tightens,* U.S. Department of Agriculture, Oct. 15, available at http://usda.mannlib.cornell.edu/usda/ers/OCS//2000s/2007/OCS-10-15-2007.pdf, accessed Nov. 2007.

Association of European Plastic Industry, 2005, *Eco-Profiles of the European Plastic Industry: Propylene, Chlorine, and Sodium Hydroxide,* available at http://lca. plasticseurope.org/index.htm, accessed Nov. 2007

Bender, M., 1999, "Economic Feasibility Review for Community-Scale Farmer Cooperatives," *Bioresource Technology,* 70: 81–87.

Britzman, D.G., 2000, *Soybean Meal — An Excellent Protein Source for Poultry Feeds,* American Soybean Association Technical Bulletin, available at http://www.asaim-europe.org/ pdf/Britzman.pdf, accessed Nov. 2007.

CETC (CANMET Energy Technology Centre), undated, *SuperCetane Technology,* available at http://canren.gc.ca/app/filerepository/381B2685235D4C6E924E0665F0A84344.pdf, accessed Nov. 2007.

Crutzen, P.J., A.R. Mosier, K.A. Smith, and W. Miniwarter, 2007, "N2O Release from Agro-Biofuel Production Negates Global Warming Reduction by Replacing Fossil Fuels," *Atmospheric Chemistry and Physics,* 7: 11191–11205.

Greiner, E., T. Kalin, and M. Yoneyama, 2005, "Epichlorohydrin," in *Chemical Economics Handbook 2004,* Report #642.3000A, SRI Consulting, Menlo Park, Calif.

EIA (Energy Information Administration), 2007a, *Annual Energy Outlook 2007,* U.S. Department of Energy, available at http://www.eia.doe.gov/oiaf/aeo/, accessed Nov. 2007.

EIA, 2007b, *Annual Energy Review 2007,* U.S. Department of Energy, available at http://www.eia.doe.gov/emeu/aer/contents.html, accessed Nov. 2007.

Haas, M.J., A.J. McAloon, W.C. Yee, et al., 2006, "A Process Model to Estimate Biodiesel Production Costs," *Bioresource Technology,* 97: 671–678.

Hill, J., E. Nelson, D. Tilman, et al., 2006, "Environmental, Economic, and Energetic Costs and Benefits of Biodiesel and Ethanol Biofuels," in *Proceedings of the National Academy of Sciences of the United States of America,* doi:10.1073/pnas.0604600103.

Intergovernmental Panel on Climate Change (IPCC), 2006, "N2O Emissions from Managed Soils, and CO2 Emissions from Lime and Urea Application," in *2006 IPCC Guidelines for National Greenhouse Gas Inventories,* Volume 4, Chapter 11, available at http://www.ipccnggip.iges.or.jp/public/2006gl/pdf/4_Volume4/V4_11_Ch11_N2O& CO2.pdf, accessed Nov. 2007.

IPCC, 1996, *Revised IPCC Guidelines for National Greenhouse Gas Inventories: Workbook,* London, U.K., 2.15–2.16.

Kalnes, T., T. Marker, and D.R. Shonnard, 2007, "Green Diesel: A Second Generation Biofuel," *International Journal of Chemical Reactor Engineering,* 5, A48.

Keller, G., M. Mintz, C. Saricks, M. Wang, and H. Ng, 2007, *Acceptance of Biodiesel as a Clean-Burning Fuel: A Draft Report in Response to Section 1823 of the Energy Policy*

Act of 2005, Center for Transportation Research, Argonne National Laboratory, prepared for Office of FreedomCAR and Vehicle Technologies, U.S. Department of Energy, Oct.

Maier, D., J. Reising, J. Briggs, K. Day, and E.P. Christmas, 1998, *High-Value Soybean Composition,* Fact Sheet #39, Grain Quality Task Force, Purdue University, available at http://www.ces.purdue.edu/extmedia/GQ/GQ-39.html, accessed Nov. 2007.

Malça, J., and F. Freire, 2006, *A Comparative Assessment of Rapeseed Oil and Biodiesel (RME) to Replace Petroleum Diesel Use in Transportation,* presented at Bioenergy I: From Concept to Commercial Processes, Tomar, Portugal, March 5–10, available at http://services.bepress.com/ cgi/viewcontent.cgi?article=1029&context=eci/bioenergy_i, accessed Nov. 2007.

Malveda, M., M. Blagoev, R. Gubler, and K. Yagi, 2005, "Glycerin," in *Chemical Economics Handbook 2004,* Report #662.5000A, SRI Consulting, Menlo Park, Calif.

National Biodiesel Board, 2007, *FAQs: How Much Biodiesel Has Been Sold in the U.S.?* available at http://www.biodiesel accessed Nov. 2007.

NRCan (Natural Resources Canada), 2003, *Technologies and Applications: The CETC SuperCetane Technology,* last updated Aug. 11, available at http://www.canren.gc.ca/ tech_appl/index.asp?CaID=2&PgId=1083, accessed Nov. 2007.

Pimentel, D., and T.W. Patzek, 2005, "Ethanol Production Using Corn, Switchgrass, and Wood; Biomass Production Using Soybean and Sunflower," *Natural Resources Research,* 14(1): 65–75.

Reuters News, 2007, *Galp Energia Selects UOP/Eni Ecofining[TM] Technology to Produce Green Diesel Fuel,* available at http://www.reuters.com/article/pressRelease/ idUS44541 +28-Nov2007+BW20071128, accessed Nov. 2007

Scharmer, K., and G. Gosse, 1996, "Ecological Impact of Biodiesel Production and Use in Europe," *The Liquid Biofuels Newsletter*-7, available at http://www.blt.bmlf.gv.at/vero liquid_biofuels_newsletter/Liquid_biofuels_Newsletter-07_e.pdf, accessed Nov. 2007.

(S&T)[2] Consultants Inc., 2004, *The Addition of NRCan's Supercetane and ROBYS[TM] Processes to GHGenius,* prepared for Natural Resources Canada.

Sheehan, J., V. Camobreco, J. Duffield, et al., 1998, *Life-Cycle Inventory of Biodiesel and Petroleum Diesel for Use in an Urban Bus,* prepared for U.S. Department of Energy, Office of Fuels Development.

UOP, 2005, *Opportunities for Biorenewables in Oil Refineries,* Final Technical Report, prepared for U.S. Department of Energy.

USDA (U.S. Department of Agriculture), 2007a, *Quick Stats: Agricultural Statistics Data Base,* available at http://www.nass.usda.gov/QuickStats/, accessed Nov. 2007.

USDA, 2007b, *Data Sets: Commodity Costs and Returns,* available at http://www.ers. usda.gov/ Data/CostsAndReturns/Fuelbystate.xls, accessed Nov. 2007.

USDA, 2007c, *Data Sets: U.S. Fertilizer Use and Price,* available at http://www.ers. usda.gov/ Data/FertilizerUse/, accessed Nov. 2007.

Wang, M.Q., 1999, *GREET 1.5 — Transportation Fuel-Cycle Model, Volume 1: Methodology, Development, Use, and Results, Volume 1,* ANL/ESD-39, Center for Transportation Research, Argonne National Laboratory, Argonne, Ill., Aug.

Wang, M., C. Saricks, and H. Lee, 2003, *Fuel-Cycle Energy and Emission Impacts of Ethanol-Diesel Blends in Urban Buses and Farming Tractors,* prepared for Illinois Department of Commerce and Economic Opportunities, by Center for Transportation Research, Argonne National Laboratory, Argonne, Ill., July.

APPENDIX 1:
ASPEN SIMULATION PROCESS
OF RENEWABLE DIESEL I (SUPER CETANE)

Victoria Putsche
Center for Transportation Technologies and Systems
National Renewable Energy Laboratory[1]

A1-1. Introduction

A preliminary analysis was conducted for a hydrogenation-derived renewable diesel (HDRD) facility on the basis of the Natural Resources Canada (NRCan) process [(S&T)[2] Consultants 2004]. NRCan has named its renewable diesel "SuperCetane." Material and energy balances were developed by using ASPEN Plus® 12.1 (super_cetane2.inp). The overall goal of the study was to confirm the preliminary overall material and energy balances provided by NRCan [(S&T)[2] Consultants 2004] and to provide input for a life-cycle analysis (LCA). The following report summarizes the basis for the analysis and its results.

A1-2. Design Basis and Process Description

HDRD is made from reacting hydrogen with oil or grease in a refinery-hydrotreating process. Several reactions occur in the conversion including hydrocracking, hydrotreating, and hydrogenation [(S&T)[2] Consultants 2004]. A commercial refinery catalyst is used to facilitate conversion.

For this analysis, the production of HDRD is based on the NRCan process, which involves hydrogen production, hydrogenation, water separation, distillation gas recycle, and steam generation. All of the unit operations were modeled except hydrogen production. It is assumed that hydrogen is supplied by an off-site hydrogen plant. Figure A1-1 is a block flow diagram of the NRCan process.

One of the important characteristics of the process is that energy demands, except electricity, are met on site. That is, a portion of the fuel gas product is combusted on site to generate steam for the process. The remaining fuel gas as well as the heavy waxy fraction are sent off site and assumed to be used for fuel. For this process configuration, the LCA will determine the emissions from the off-site fuel gas and heavies combustion as well as the electricity generation and will apportion it appropriately to the main process. This analysis will estimate the emissions from the fuel gas combusted on site.

---

[1] Contact person for further information: Paul Bergeron (Paul_Bergeron@nrel.gov) of National Renewable Energy Laboratory.

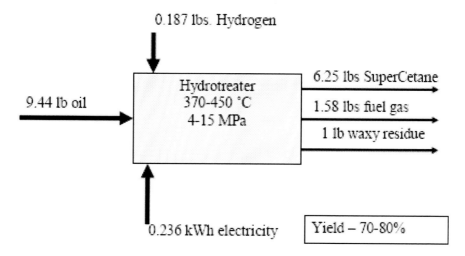

Figure A1-1. HDRD (SuperCetane) Block Flow Diagram.

The renewable diesel process was modeled by using numerous assumptions and data sources. Table A1-1 summarizes the key design parameters and their sources.

Table A-1. Design Basis

Parameter	Value	Source
Feedstock		
Type	Soybean oil	Most common oil in U.S. for biodiesel
Throughput	100 lb/h	For LCA analysis
Feedstock fatty acid composition		
(wt fraction)		
Linolenic acid	0.075	
Palmitic acid	0.11	
Stearic acid	0.041	
Oleic acid	0.22	
Linoleic acid	0.54	
Arachidic acid	0.014	
Hydrogenation design		
Temperature	325 °C	
Pressure	500 psia	
Yields (per pound inlet feed) SuperCetane		Derived from published yields [(S&T)2 Consultants 2004]
Water	64.9%	
CO2	5.0%	
Propane	8.2%	
Hydrogen	8.2%	
Naphtha	10.4% 0.35	

Several of these assumptions, particularly the feedstock choice and facility size, require further explanation. The feedstock selected was soybean oil, even though many of the feedstocks in the literature were rapeseed oil or other oils, because it is the most prevalent oil in fuels production (i.e., biodiesel), and one of the purposes of the study was to compare the environmental impacts of HDRD to biodiesel, and the most thorough LCA of biodiesel (Sheehan et al. 1998) was based on soybean oil. The facility size of 100 lb/h was selected as an easy, round number for the LCA. The results of most LCAs are shown on a pound of feed or product basis since the impacts are directly scalable to throughput. Therefore, this simple number was selected, even though this would not be a typical facility size.

A1-3. Model Description

An ASPEN Plus® model (super_cetane) was developed for the NRCan SuperCetane process, based largely on the (S&T)² report [(S&T)² 2004]. ASPEN Plus® is a steady-state process simulator, and Appendix A1-6 contains the input file for the model.

The ASPEN Plus® HDRD model has one flowsheet to model the four major process areas: hydrogenation, sour water separation, stripping, and pressure swing adsorption (PSA)/gas recycle. Each of these areas is briefly discussed, and the flow diagram from ASPEN Plus® is presented. The flow diagram shows only those unit operations modeled in ASPEN Plus®. Equipment used for operations such as conveyance, size reduction, and storage is generally not included in the model. The power requirements of this equipment, however, are included and are modeled as work streams.

ASPEN Plus® is composed of physical property and unit operation models that are combined into a process model. The simulation can be broken into three major sections: components (i.e., chemical species), physical property option sets (e.g., what set of physical property models to use), and the flowsheet (i.e., the series of unit operations). Each of these sections is described in more detail below.

Components

Fourteen components were modeled in the simulation; all were modeled as conventional (e.g., water) components in the mixed substream. The following is a list of the components in the simulation:

- Hydrogen – H_2
- Linolenic acid – $C_{18}H_{30}O_2$
- Palmitic acid – $C_{16}H_{32}O_2$
- Stearic acid – $C_{18}H_{36}O_2$
- Linoleic acid – $C_{18}H_{32}O_2$
- Arachidic acid – $C_{20}H_{40}O_2$
- Oleic acid – $C_{18}H_{34}O_2$
- Green Diesel – $C_{18}H_{38}$
- Water – H_2O
- Hydrogen Sulfide – H_2S
- Ammonia – NH_3

- Propane – C3H8
- Naphtha
- Oxygen – O2
- Nitrogen – N2
- Wax – C26H54
- Carbon dioxide – CO2

Green diesel is not a specific compound but is a complex mixture of hydrocarbons; however, for simplicity, it was modeled as a single component, c_{18H38}, which is within the range of diesel hydrocarbons. Green diesel was specified with a specific gravity of 0.78 (Marker, T. 2007) and a MW of 254. Naphtha was specified with a specific gravity of 0.7 and a MW of 100.

As noted earlier, the vegetable oil feed was modeled as a mixture of six fatty acids: linolenic acid, palmitic acid, stearic acid, linoleic acid, arachidic acid, and oleic acid. All of these components are available in the ASPEN Plus® databanks. Table A1-2 shows the molecular formula, the component name in the model, and the weight fraction in the feed of each fatty acid.

Table A1-2. Organic Acid Composition of Bio-Oil

Organic Fatty Acid	Composition	Component Name	Weight Fraction
Linolenic	C18H30O2	LINOL3	0.075
Palmitic	C16H32O2	PALM	0.11
Stearic	C18H36O2	STEARIC	0.041
Oleic	C18H34O2	OLEIC	0.22
Linoleic	C18H32O2	LINOL2	0.54
Arachidic	C20H40O2	ARACHID	0.014

One Henry component, CO2, was specified. The Henry's constants were obtained from ASPEN Plus®.

Physical Property Option Sets

The physical property set selected was POLYUF with properties estimated by using the POLYNRTL method. Physical property databanks used in the simulation were PURE13, AQUEOUS, SOLIDS AND INORGANIC.

Flowsheet

One flowsheet was developed for the process: (A1000). The flowsheet is briefly discussed, and flow diagrams from ASPEN Plus® are presented. The flow diagrams (Figure A1-2) show only those unit operations modeled in ASPEN Plus®. Equipment used for operations such as conveyance and storage are generally not included in the model and are thus not shown. Similarly, certain complex unit operations (e.g., gas turbine) require several ASPEN Plus® models (e.g., compressors, reactors, heat exchangers).

Bio-oil is introduced into the process in stream 101. It is assumed to be at ambient conditions (i.e., 68 F and 14.7 psia) with a flow rate of a nominal 100 lb/h. The 100 lb/h value was selected as it would be easily scaled to any other value; since the model was developed to be the basis for an LCA, any flow rate would be reasonable.

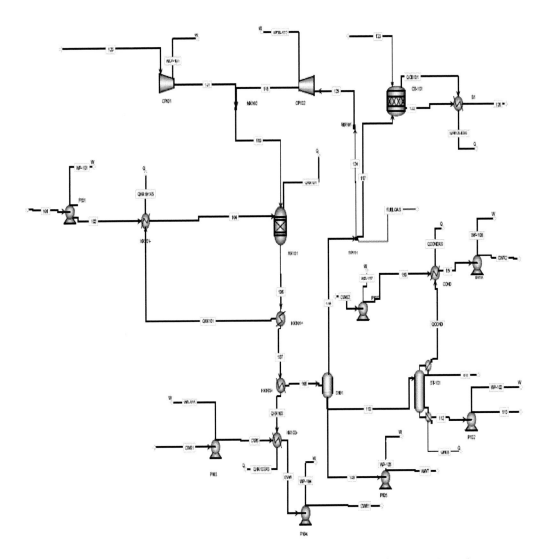

Figure A1-2. ASPEN Simulation Process Flowcharts for Renewable Diesel I (SuperCetane).

Hydrogenation

As shown in Figure A1-2, the soybean oil feed (Stream 101) is pumped to 500 psia (P-101) and then mixed with recycle oil (Stream 110C) from the splitter, SP-101, following the sour water separator (S101). This stream is then heated to 290°F by exchange with hydrogenator effluent (Stream 106) in 11X101+ and 11X101−. The next stage of the process is the hydrogenator, where the oil stream is combined with the inlet hydrogen (Stream 120) and recycle fuel gas (Stream 118) and reacted.

The hydrogenator (RX101) is modeled as an RYIELD reactor. All of the incoming oil is converted to gas (e.g., CO2, 112, propane), water, green diesel (GDSL), waxes, and a small amount of naphtha. As noted in the design basis, the yield of green diesel is estimated at 64.5% of the total inlet feed streams on a mass basis. The hydrogenation reactions are exothermic, and there is excess heat (QRX101) after the reactor is brought to reaction temperature (325°C).

After the oil feed is preheated, the hydrogenator effluent (Stream 107) cooled with water (cooling Stream CWS1) to 100°F in 11X103. The cooled reactor products are then sent to the sour water separation, S101.

Sour Water Separation

In sour water separation, the gases (Stream 115) are flashed off and sent to a splitter (SP101) for recycle, combustion, and product recovery. The aqueous stream is decanted and sent to wastewater treatment (Stream 109). After the separator, the organic stream (110) is sent to a distillation column (ST-101) for product recovery.

Product Recovery

The product recovery area consists of a distillation column where the SuperCetane (Stream 111) with a small amount of naphtha is separated from the heavies (Stream 112). The distillation column is modeled as a RADFRAC column with eight stages with both a condenser and a reboiler. The system is operated at 100 psi (stage 1). The feed is introduced on stage 5, while SuperCetane is recovered on stage 1, and the heavies are taken off on stage 8.

Gas Recycle

As noted earlier, the off-gas from the water separator, S101, is sent to a splitter where it is separated for gas recycle (124), combustion (117), and product (FUELGAS). The amount of product is controlled by overall process yields, while the amount sent to combustion is specified so that the system's energy demand is satisfied.

Heat Generation

The last major section of the flowsheet is steam generation. Here, some of the fuel gas is combusted (CB-101), which is operated at 1700 F. Heat is recovered from the off-gases in a HEATER block, B1. The amount of heat recovery is compared to the process heat demands [e.g., the reboiler (QREB)] to ensure that enough heat is available. A more rigorous model could be developed that would generate steam and meet the specific heat demands of each unit operation. For this analysis, this gross heat balance was deemed sufficient.

A1-4 Results and Discussion

This effort was aimed at confirming the material and energy balances summarized for the NRCan process. As shown in the table below, the ASPEN Plus® model shows good agreement with the published literature. All of the yields and utility requirements are similar between the model and the literature. Table A1-3 compares the results of this modeling effort and the values from the (S&T)[2] Consultants (2004).

Table A1-3. Comparison of Overall Mass and Energy Balances

Feedstock	NRCan Yield per 100 lb oil	Current Analysis per 100 lb Oil
Oil	100	100
H2	1.98	1.98
Air		63.47
Products		
Fuel Gas	16.74	16.74
HDRD	66.21	66.2
Naphtha	0.36	
Heavies (113)	11.60	11.6
Waste water		4.39
Flue Gas (lb/h)		66.52
Flue Gas (126) (scf)		24.2
Utilities		
Electricity (kWh)	2.50	2.61
Cooling water (lb/h)		4307

Besides SuperCetane, this process generates three other products: fuel gas, naphtha, and heavies. The amount of naphtha is very small and is included in the SuperCetane product. Table A1-4 summarizes the calculated compositions of the other products.

Table A1-4. Product Compositions

Product	Composition
Fuel gas	
Propane	25.45
Carbon dioxide	27.96
Water	2.33
Hydrogen	43.68
Naphtha	0.58
LHV (Btu/lb)	27,999
Heavies	
Wax	80
Naphtha	20
LHV (Btu/lb)	20,617

In addition to the material and energy balance, the analysis projected the air emissions from the process. As noted earlier, it is assumed that a portion of the fuel gas, which is primarily propane, is combusted to make steam to meet the energy demand of the process. Air emissions of criteria pollutants were estimated on the basis of the U.S. Environmental Protection Agency's AP-42 emission factors. The fuel gas is a mix of several gases, but for this analysis, the emissions were assumed to be equivalent to natural gas combustion. Table A1-5 summarizes the emission factors and the emission rate of each pollutant.

Table A1-5. Air Emission Factors

Pollutant	Emission Factors (lb/MM scf fuel)	Emissions (lb/100 lb product)
CO	84	3.07E-03
NO_x	32	1.17E-03
PM	7.6	2.78E-04
VOCs	5.5	2.01E-04

The NRCan process uses hydrogenation to convert bio-oils like soybean oil into a diesel substitute. Several companies are looking into this process. This analysis developed an ASPEN Plus® model of the process and compared its results with published results by $(S\&T)^2$ Consultants (2004). Good agreement was obtained between the two studies. These results will be used to develop an LCA for this process.

A1-5. References

Sheehan, J., V. Camobreco, J. Duffield, M. Graboski, H. Shapouri. 1998, *An Overview of Biodiesel and Petroleum Diesel Life Cycles*, joint study sponsored by U.S. Department of Agriculture and U.S. Department of Energy, NREL/TP-580-24772.

$(S\&T)^2$ Consultants Inc., 2004. *The Addition of NRCan's SuperCetane and ROBYS ᵀᴹ Processes to GHGenius*, prepared for Natural Resources Canada, March 30.

A1-6. ASPEN Plus® Input File: Super_Cetane2.inp

```
;Input Summary created by Aspen Plus Rel. 20.0 at 16:58:40 Sun Oct 21, 2007
;Directory E:\HDRD Filename E:\HDRD\super_cetane2.inp
;
TITLE 'Super Cetane'
IN-UNITS ENG DENSITY='lb/gal' POWER=kW VOLUME=gal & MOLE-DENSITY
='lbmol/gal' MASS-DENSITY='lb/gal'

DEF-STREAMS CONVEN ALL

DATABANKS PURE13 / AQUEOUS / SOLIDS / INORGANIC / & NOASPENPCD

PROP-SOURCES PURE13 / AQUEOUS / SOLIDS / INORGANIC

COMPONENTS
H2 H2 /
LINOL3 C18H30O2 / PALM C16H32O2 / STEARIC C18H36O2 / OLEIC C18H34O2 /
LINOL2 C18H32O2 / ARACHID C20H40O2 / GDSL C18H38 /
H2O H2O / H2S H2S / NH3 H3N / PROPANE C3H8 /
NAPTHA / CO2 CO2 / WAX C26H54 /
O2 O2 /
N2 N2
```

PC-USER
IN-UNITS ENG
PC-DEF ASPEN GDSL GRAV=0.749 MW=254.PC-DEF ASPEN NAPTHA
GRAV=0.7 MW=72.

ADA-SETUP
ADA-SETUP PROCEDURE=REL9

HENRY-COMPS HC-1 CO2

FLOWSHEET
BLOCK RX101 IN=119 104 OUT=106 QRX101 BLOCK S101 IN=108 OUT=115 110
109 BLOCK P101 IN=101 OUT=102 WP-101 BLOCK P102 IN=112 OUT=113 WP-
102 BLOCK CP102 IN=125 OUT=118 WCP-102 BLOCK CP101 IN=120 OUT=121
WCP-101 BLOCK HX101+ IN=106 OUT=107 QHX101
BLOCK HX101- IN=102 QHX101 OUT=104 QHX101XS BLOCK MX102 IN=118 121
OUT⁻119
BLOCK ST-101 IN=110 OUT=111 112 QCOND QREB BLOCK HX103+ IN=107
OUT=108 QHX103
BLOCK HX103- IN=CWS QHX103 OUT=CWR QHX103XS BLOCK P105 IN=109
OUT=WWT WP-105
BLOCK P103 IN=CWS1 OUT=CWS WP-103 BLOCK P104 IN=CWR OUT=CWR1
WP-104 BLOCK SP101 IN=115 OUT=117 FUELGAS 124 BLOCK CB-101 IN=117
123 OUT=122 QCB101 BLOCK MX101 IN=124 OUT=125
BLOCK COND IN=150 QCOND OUT=151 QCONDXS
BLOCK P107 IN=CWS2 OUT=150 WP-107 BLOCK P108 IN=151 OUT=CWR2 WP-
108 BLOCK B1 IN=122 QCB101 OUT=126 QPROCESS

PROPERTIES POLYUF HENRY-COMPS=HC-1 PROPERTIES POLYNRTL

PROP-DATA HENRY-1
IN-UNITS ENG
PROP-LIST HENRY
BPVAL CO2 H2O 175.2762325 -15734.78987 -21.66900000 &
6.12550005E-4 31.73000375 175.7300026 0.0

STREAM 101
IN-UNITS ENG
SUBSTREAM MIXED TEMP=68. PRES=14.7 MASS-FLOW=100. MASS-FRAC
LINOL3 0.075 / PALM 0.11 / STEARIC 0.041 / & OLEIC 0.22 / LINOL2 0.54 /
ARACHID 0.014

STREAM 117
IN-UNITS ENG
SUBSTREAM MIXED TEMP=68. PRES=14.7 MASS-FLOW=1. MASS-FRAC H2 1.

STREAM 120
IN-UNITS ENG
SUBSTREAM MIXED TEMP=68. PRES=14.7 MASS-FLOW=100. MASS-FRAC H2
1.

STREAM 123
SUBSTREAM MIXED TEMP=68. PRES=14.7 MASS-FLOW=67. MOLE-FRAC O2
0.21 / N2 0.79

STREAM 125
SUBSTREAM MIXED TEMP=68. PRES=14.7 MASS-FLOW=10. MASS-FRAC H2 1.

STREAM CWS
IN-UNITS ENG
SUBSTREAM MIXED TEMP=35. <C> PRES=500. MASS-FLOW=100.

MASS-FRAC H2O 1.

STREAM CWS1
IN-UNITS ENG
SUBSTREAM MIXED TEMP=35. <C> PRES=14.7 MASS-FLOW=100. MASS-FRAC
H2O 1.

STREAM CWS2
SUBSTREAM MIXED TEMP=35. <C> PRES=14.7 MASS-FLOW=100. MASS-FRAC
H2O 1.

DEF-STREAMS HEAT QCB101

DEF-STREAMS HEAT QCOND

DEF-STREAMS HEAT QCONDXS

DEF-STREAMS HEAT QHX101

DEF-STREAMS HEAT QHX101XS

DEF-STREAMS HEAT QHX103

DEF-STREAMS HEAT QHX103XS

DEF-STREAMS HEAT QPROCESS

DEF-STREAMS HEAT QREB

DEF-STREAMS HEAT QRX101

DEF-STREAMS WORK WCP-101

DEF-STREAMS WORK WCP-102

DEF-STREAMS WORK WP-101

DEF-STREAMS WORK WP-102

DEF-STREAMS WORK WP-103

DEF-STREAMS WORK WP-104 DEF-STREAMS WORK WP-105 DEF-STREAMS WORK WP-107 DEF-STREAMS WORK WP-108

BLOCK MX101 MIXER

BLOCK MX102 MIXER IN-UNITS ENG

BLOCK SP101 FSPLIT FRAC FUELGAS 0.5 MASS-FLOW 124 10.

BLOCK B1 HEATER
PARAM TEMP=100. PRES=14.7

BLOCK COND HEATER PARAM PRES=14.7 DELT=15.

BLOCK HX101+ HEATER
IN-UNITS ENG
PARAM TEMP=110. PRES=500.

BLOCK HX101- HEATER
IN-UNITS ENG
PARAM TEMP=567. PRES=500.

BLOCK HX103+ HEATER
IN-UNITS ENG
PARAM TEMP=100. PRES=500.

BLOCK HX103- HEATER
IN-UNITS ENG
PARAM PRES=500. DELT=15.

BLOCK S101 FLASH2
IN-UNITS ENG
PARAM TEMP=100. PRES=175.

BLOCK-OPTION FREE-WATER=YES

BLOCK ST-101 RADFRAC
IN-UNITS ENG
PARAM NSTAGE=8
COL-CONFIG CONDENSER=TOTAL REBOILER=KETTLE FEEDS 110 5
PRODUCTS 111 1 L / 112 8 L
PRODUCTS QREB 8 / QCOND 1
P-SPEC 1 100.
COL-SPECS DP-STAGE=1. MASS-D=66.2 MOLE-RR=0.1

BLOCK CB-101 RSTOIC
PARAM TEMP=1700. PRES=0. COMBUSTION=YES PROD-NOX=NO2 STOIC 1
MIXED H2 -1. / O2 -0.5 / H2O 1.
STOIC 2 MIXED PROPANE -1. / O2 -5. / CO2 3. / H2O 4.
STOIC 3 MIXED NAPTHA -1. / O2 -8. / CO2 5. / H2O 6.
CONV 1 MIXED H2 1.
CONV 2 MIXED PROPANE 1.
CONV 3 MIXED NAPTHA 1.

BLOCK RX101 RYIELD
IN-UNITS ENG
PARAM TEMP=325. <C> PRES=500.
MASS-YIELD MIXED GDSL 0.8415 / H2O 0.02125 / CO2 & 0.10625 / PROPANE
0.029 / H2 0.001 / NAPTHA 0.01 / & WAX 0.104

BLOCK P101 PUMP
IN-UNITS ENG
PARAM PRES=500.

BLOCK P102 PUMP
IN-UNITS ENG
PARAM DELP=10.

BLOCK P103 PUMP
IN-UNITS ENG
PARAM PRES=500. PUMP-TYPE=TURBINE

BLOCK P104 PUMP
IN-UNITS ENG
PARAM DELP=10.

BLOCK P105 PUMP
IN-UNITS ENG
PARAM DELP=10.

BLOCK P107 PUMP PARAM DELP=10.

BLOCK P108 PUMP PARAM DELP=10.

BLOCK CP101 COMPR
IN-UNITS ENG
PARAM TYPE=ISENTROPIC PRES=500. MODEL-TYPE=TURBINE

BLOCK CP102 COMPR
IN-UNITS ENG
PARAM TYPE=ISENTROPIC PRES=500. MODEL-TYPE=TURBINE

DESIGN-SPEC COMBAIR
DEFINE O2OUT MASS-FLOW STREAM=122 SUBSTREAM=MIXED &
COMPONENT=O2
DEFINE O2IN MASS-FLOW STREAM=123 SUBSTREAM=MIXED &
COMPONENT−O2
SPEC "O2IN" TO "11*O2OUT"
TOL-SPEC "1"
VARY   STREAM-VAR   STREAM=123   SUBSTREAM=MIXED   &
VARIABLE=MASS-FLOW
LIMITS "50" "150"

DESIGN-SPEC DS-FGAS
DEFINE   SPLT   BLOCK-VAR   BLOCK=SP101   SENTENCE=FRAC   &
VARIABLE=FRAC ID1=FUELGAS
DEFINE FGAS STREAM-VAR STREAM=FUELGAS SUBSTREAM=MIXED &
VARIABLE=MASS-FLOW
SPEC "FGAS" TO "16.74"
TOL-SPEC "0.05"
VARY BLOCK-VAR BLOCK=SP101 SENTENCE=FRAC VARIABLE=FRAC &
ID1=FUELGAS
LIMITS "0.05" "0.95"

DESIGN-SPEC DS-HX101
IN-UNITS ENG
DEFINE   QXS   INFO-VAR   INFO=HEAT   VARIABLE=DUTY   &
STREAM=QHX101XS
SPEC "QXS" TO "0.0"
TOL-SPEC "0.1"
VARY BLOCK-VAR BLOCK=HX101+ VARIABLE=TEMP SENTENCE=PARAM
LIMITS "100" "617"

DESIGN-SPEC DS-HX103
IN-UNITS ENG
DEFINE CWIN STREAM-VAR STREAM=CWS1 SUBSTREAM=MIXED &

VARIABLE=MASS-FLOW
DEFINE QXS INFO-VAR INFO=HEAT VARIABLE=DUTY &
STREAM=QHX103XS SPEC "QXS" TO "0"
TOL-SPEC "0.1"
VARY   STREAM-VAR   STREAM=CWS1   SUBSTREAM=MIXED   &
VARIABLE=MASS-FLOW
LIMITS "100" "10000"

DESIGN-SPEC DS-QCOND
DEFINE QXS INFO-VAR INFO=HEAT VARIABLE=DUTY STREAM=QCONDXS
DEFINE CWIN STREAM-VAR STREAM=CWS2 SUBSTREAM=MIXED &
VARIABLE=MASS-FLOW SPEC "QXS" TO "0"
TOL-SPEC "0.1"
VARY   STREAM-VAR   STREAM=CWS2   SUBSTREAM=MIXED   &
VARIABLE=MASS-FLOW
LIMITS "5" "5000"

EO-CONV-OPTI

CALCULATOR H2IN
IN-UNITS ENG
DEFINE H2IN STREAM-VAR STREAM=120 SUBSTREAM=MIXED &
VARIABLE=MASS-FLOW
DEFINE OILIN STREAM-VAR STREAM=101 SUBSTREAM=MIXED &
VARIABLE=MASS-FLOW
F H2IN = 0.0198*OILIN
READ-VARS OILIN

CALCULATOR HYDCRK
IN-UNITS ENG
DEFINE   FEED   STREAM-VAR   STREAM=101   SUBSTREAM=MIXED   &
VARIABLE=MASS-FLOW
DEFINE   GDYLD   BLOCK-VAR   BLOCK=RX101   VARIABLE=YIELD   &
SENTENCE=MASS-YIELD ID1=MIXED ID2=GDSL
DEFINE   PROYLD   BLOCK-VAR   BLOCK=RX101   VARIABLE=YIELD   &
SENTENCE=MASS-YIELD ID1=MIXED ID2=PROPANE
DEFINE   CO2YLD   BLOCK-VAR   BLOCK=RX101   VARIABLE=YIELD   &
SENTENCE=MASS-YIELD ID1=MIXED ID2=CO2
DEFINE   H2OYLD   BLOCK-VAR   BLOCK=RX101   VARIABLE=YIELD   &
SENTENCE=MASS-YIELD ID1=MIXED ID2=H2O
DEFINE   FD105   STREAM-VAR   STREAM=104   SUBSTREAM=MIXED   &
VARIABLE=MASS-FLOW
DEFINE   FD119   STREAM-VAR   STREAM=119   SUBSTREAM=MIXED   &
VARIABLE=MASS-FLOW
DEFINE   H2IN   STREAM-VAR   STREAM=120   SUBSTREAM=MIXED   &
VARIABLE=MASS-FLOW

```
DEFINE   H2YLD   BLOCK-VAR   BLOCK=RX101   VARIABLE=YIELD   &
SENTENCE=MASS-YIELD ID1=MIXED ID2=H2
DEFINE   NPYLD   BLOCK-VAR   BLOCK=RX101   VARIABLE=YIELD   &
SENTENCE=MASS-YIELD ID1=MIXED ID2=NAPTHA
DEFINE FGAS STREAM-VAR STREAM=FUELGAS SUBSTREAM=MIXED &
VARIABLE=MASS-FLOW
DEFINE   WXYLD   BLOCK-VAR   BLOCK=RX101   VARIABLE=YIELD   &
SENTENCE=MASS-YIELD ID1=MIXED ID2=WAX
F TTLFD = FD105+FD119
F GDYLD = 0.649*(FEED+H2IN)/TTLFD
F PROYLD = 0.082*(FEED+H2IN)/TTLFD
F CO2YLD = 0.082*(FEED+H2IN)/TTLFD
F H2OYLD = 0.050*(FEED+H2IN)/TTLFD
F NPYLD = 0.0035*(FEED+H2IN)/TTLFD F WXYLD = 0.104*(FEED+H2IN)/TTLFD
F SUM = GDYLD+PROYLD+CO2YLD+H2OYLD+NPYLD+WXYLD
F DIFF = TTLFD - (SUM*TTLFD)
F H2YLD = DIFF/TTLFD
F WRITE(NHSTRY,*)SUM,DIFF,H2YLD READ-VARS FEED FD105 FD119 H2IN
FGAS
WRITE-VARS GDYLD PROYLD CO2YLD H2OYLD H2YLD NPYLD WXYLD
BLOCK-OPTION SIM-LEVEL=4

STREAM-REPOR NOMOLEFLOW MASSFLOW

PROPERTY-REP NOPARAM-PLUS
```

APPENDIX 2:
ASPEN SIMULATION PROCESS OF RENEWABLE DIESEL II
(HYDROGENATION-DERIVED RENEWABLE DIESEL)

Victoria Putsche
Center for Transportation Technologies and Systems
National Renewable Energy Laboratory[2]

A2-1. Introduction

A preliminary analysis was conducted for a hydrogenation-derived renewable diesel (HDRD) facility on the basis of the UOP process (UOP 2006). Material and energy balances were developed by using ASPEN Plus® 12.1 (uop_hdrd.inp). The overall goal of the study was to confirm the preliminary overall material and energy balances provided by UOP (UOP

[2] Contact person for further information: Paul Bergeron (Paul_Bergeron@nrel.gov) of National Renewable Energy Laboratory.

2006; Markel 2006) and to provide input for a life-cycle analysis (LCA). The following report summarizes the basis for the analysis and its results.

A2-2. Design Basis and Process Description

HDRD is made from reacting hydrogen with oil or grease in a refinery-hydrotreating process. Two primary reactions occur in the conversion: hydrodeoxygenation and decarboxylation (UOP 2006)
Hydrodeoxygenation:

C_nCOOH (bio-oil) + 3 H2 → Cn+1 (HDRD) + 2 H2O
Decarboxylation:
C_nCOOH (bio-oil) → C_n (HDRD) + CO2

The selectivity of the reactions depends on the processing conditions.
For this analysis, the production of HDRD is based on the UOP process, which is composed of hydrogen production, hydrogenation, separation, distillation, and pressure swing adsorption (PSA). All of the unit operations were modeled except hydrogen production. It is assumed that hydrogen is supplied by a hydrogen plant. Figure A2-1 is a block flow diagram of the HDRD process.

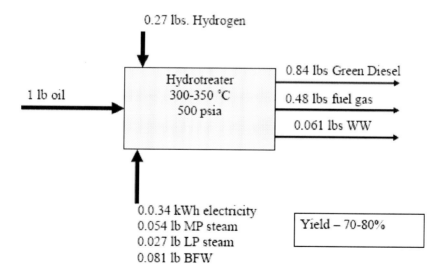

Figure A2-1 HDRD Block Flow Diagram.

One of the important characteristics of the process is that energy demands are met off site. That is, the fuel gas product is not combusted on site to generate steam for the process; it is assumed that steam is sent to the process from an off-site source. Similarly, the process also generates a fuel gas, which is also sent off site and used for fuel. For this process configuration, the LCA will determine the emissions from the fuel gas combustion as well as the steam and electricity generation and will apportion it appropriately to the main process.

The renewable diesel process was modeled by using numerous assumptions and data sources. Table A2-1 summarizes the key design parameters and their sources.

Several of these assumptions, particularly the feedstock choice and facility size, require further explanation. The feedstock selected was soybean oil, even though many of the feedstocks in the literature were rapeseed oil or other oils, because it is the most prevalent oil in fuels production (i.e., biodiesel) and because one of the purposes of the study was to compare the environmental impacts of HDRD to biodiesel, and the most thorough LCA of biodiesel (Sheehan et al. 1998) was based on soybean oil. The facility size of 100 lb/h was selected as an easy, round number for the LCA. The results of most LCAs are shown on a pound of feed or product basis, since the impacts are directly scalable to throughput. Therefore, this simple number was selected, even though this would not be a typical facility size.

Table A2-1. Design Basis

Parameter	Value	Source
Feedstock		
Type	Soybean oil	Most common oil in US for biodiesel
Throughput	100 lb/h	For LCA analysis
Feedstock fatty acid composition (wt fraction)		
Linolenic acid		
Palmitic acid	0.075	
Stearic acid	0.11	
Oleic acid	0.041	
Linoleic acid	0.22	
Arachidic acid	0.54	
	0.014	
Hydrogenation design		
Temperature	325°C	UOP 2006
Pressure	500 psia	UOP 2006
Yields (per pound inlet feed)		
HDRD		UOP 2006
Water	84.15%	
CO2	2.125%	
Propane	10.625%	
Hydrogen	2.9%	
	0.1%	

A2-3. Model Description

An ASPEN Plus® model (uop_hdrd) was developed for the pyrolysis process, largely on the basis of the UOP report (UOP 2006). ASPEN Plus® is a steady-state process simulator. Appendix A2-6 contains the input file for the model.

The ASPEN Plus® HDRD model has one flowsheet to model the four major process areas: hydrogenation, sour water separation, stripping, and pressure swing adsorption (PSA)/gas recycle. Each of these areas is briefly discussed, and the flow diagram from ASPEN Plus® is presented. The flow diagram shows only those unit operations modeled in ASPEN Plus®. Equipment used for operations such as conveyance, size reduction, and storage us generally not included in the model. The power requirements of this equipment, however, are included and are modeled as work streams.

ASPEN Plus® is composed of physical property and unit operation models that are combined into a process model. The simulation can be broken into three major sections: components (i.e., chemical species), physical property option sets (e.g., what set of physical property models to use), and the flowsheet (i.e., the series of unit operations). Each of these sections is described in more detail below.

Components

Fourteen components were modeled in the simulation; all were modeled as conventional (e.g., water) components in the mixed substream. The following is a list of the components in the simulation:

- Hydrogen – H_2
- Linolenic acid – $C_{18}H_{30}O_2$
- Palmitic acid – $C_{16}H_{32}O_2$
- Stearic acid – $C_{18}H_{36}O_2$
- Linoleic acid – $C_{18}H_{32}O_2$
- Arachidic acid – $C_{20}H_{40}O_2$
- Oleic acid – $C_{18}H_{34}O_2$
- Green Diesel – $C_{18}H_{38}$
- Water – H_2O
- Hydrogen Sulfide – H_2S
- Ammonia – NH_3
- Propane – C_3H_8
- Naptha
- Carbon dioxide – CO_2

Green diesel is not a specific compound but is a complex mixture of hydrocarbons; however, for simplicity, it was modeled as a single component, C_{18H38}, which is within the range of diesel hydrocarbons. Green diesel was specified with a specific gravity of 0.78 (Marker, T. 2007) and a MW of 254. Naphtha was specified with a specific gravity of 0.7 and a MW of 100.

As noted earlier, the vegetable oil feed was modeled as a mixture of six fatty acids: linolenic acid, palmitic acid, stearic acid, linoleic acid, arachidic acid, and oleic acid. All of these components are available in the ASPEN Plus® databanks. Table A2-2 shows the molecular formula, the component name in the model, and the weight fraction in the feed of each fatty acid.

One Henry component, CO_2, was specified. The Henry's constants were obtained from ASPEN Plus®.

Table A2-2. Organic Acid Composition of Bio-Oil

Organic Fatty Acid	Composition	Component Name	Weight Fraction
Linolenic	C18H30O2	LINOL3	0.075
Palmitic	C16H32O2	PALM	0.11
Stearic	C18H36O2	STEARIC	0.041
Oleic	C18H34O2	OLEIC	0.22
Linoleic	C18H32O2	LINOL2	0.54
Arachidic	C20H40O2	ARACHID	0.014

Physical Property Option Sets

The physical property set selected was POLYUF with properties estimated by using the POLYNRTL method. Physical property databanks used in the simulation were PURE13, AQUEOUS, SOLIDS AND INORGANIC.

Flowsheet

One flowsheet was developed for the process: (A1000). The flowsheet is briefly discussed, and flow diagrams from ASPEN Plus® are presented. The flow diagrams (Figure A2-2) show only those unit operations modeled in ASPEN Plus®. Equipment used for operations such as conveyance and storage are generally not included in the model and are thus not shown. Similarly, certain complex unit operations (e.g., gas turbine) require several ASPEN Plus® models (e.g., compressors, reactors, heat exchangers).

Bio-oil is introduced into the process in stream 101. It is assumed to be at ambient conditions (i.e., 68 F and 14.7 psia) with a flow rate of a nominal 100 lb/h. The 100-lb/h value was selected as it would be easily scaled to any other value; since the model was developed to be the basis for an LCA, any flow rate would be reasonable.

Hydrogenation

As shown in Figure A2-2, the soybean oil feed (Stream 101) is pumped to 500 psia (P-101) and then mixed with recycle oil (Stream 110C) from the splitter, SP-101, following the sour water separator (S101). This stream is then heated to 290 F by exchange with hydrogenator effluent (Stream 106) in HX101+ and HX101-. It is then further heated to 370 F with medium-pressure steam, MPSS (150 psig). The next stage of the process is the hydrogenator, where the oil stream is combined with the inlet hydrogen (Stream 119) and reacted.

The hydrogenator (RX101) is modeled as an RYIELD reactor. All of the incoming oil is converted to gas (e.g., CO2, 112, propane), water, and green diesel (GDSL). As noted in the design basis, the yield of green diesel is estimated at 84.15% of the inlet feed streams on a mass basis. The hydrogenation reactions are exothermic, and there is excess heat (QRX101) after the reactor is brought to reaction temperature (325 C).

After the oil feed is preheated the hydrogenator effluent (Stream 107) is cooled with cooling water (Stream CWS1) to 100 F in 11X103. The cooled reactor products are then sent to the sour water separation, S101.

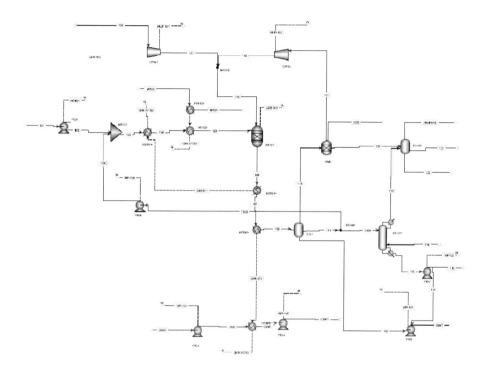

Figure A2-2. ASPEN Simulation Process Flowcharts FOR Renewable Diesel II.

Sour Water Separation

In sour water separation, the gases (Stream 115) are flashed off and sent to the PSA for recovery, and the aqueous stream is decanted and sent to wastewater treatment (Stream 109). After the separator, a portion of the organic stream (110B) is recycled to the hydrogenator inlet. The remainder (Stream 110A) is sent to a stripping column (ST-101) for product recovery.

Product Recovery

The product recovery area consists of a stripping column where LP (50 psig) steam (Stream 114) is used to remove the light ends from the green diesel product (112). The stripping column is modeled as a RADFRAC column with eight stages without a condenser or reboiler under atmospheric pressure.

The overheads are sent to the flash unit of the PSA system, FL-101. The product stream is taken from the bottom of the column (Stream 112).

Pressure Swing Adsorption (PSA)

The PSA system is a complex batch unit operation that was treated basically as a black box for this simulation. It is modeled as two unit operations in series, a separator block (PSA) followed by a flash block (FL-101). The separator block is assumed to remove all of the hydrogen in the overhead stream (111). The recovered hydrogen is then compressed (CP102) to 500 psia before introduction into the hydrogenator.

In addition to hydrogen, the PSA unit operation has two other outlet streams: CO_2 and Stream 125. The CO_2 stream contains all of the carbon dioxide from the operation and is released to the atmosphere. Stream 125 contains a mixture of water, propane, and other organics. These are separated in FL-101 modeled as a FLASH2. As shown in the diagram, FL-101 has two inlets (Streams 111 and 125) and three outlets: PROPANE and Streams 122 and 123. Stream 111 is the overheads from the stripping column, ST-101. PROPANE is a fuel gas, composed primarily of propane (93%) with small amounts of green diesel and CO_2.

A2-4. Results and Discussion

This effort was aimed at confirming the material and energy balances summarized for the UOP HDRD process as found in UOP (2006) and Markel (2006). As shown in Table A2-3, the ASPEN Plus® model shows good agreement with the published literature.

Table A2-3. Comparison of Overall Mass and Energy Balances

Feedstock	UOP Yield per 100 lb of feed	Current Analysis Yield per 100 lb of feed
Oil	100.00	100.00
H2	2.72	2.72
LP steam	2.72	2.80
Products		
Propane mix gas	4.75	5.02
HDRD	84.19	85.23
CO2		7.01
Waste water	6.11	8.27
Utilities		
Electricity (kWh)	3.39	2.34
LP Steam (into process)	2.72	2.80
MP steam	5.43	5.37
Cooling water	1,356	2,310
Boiler feed water	8.15	8.17
Total steam (Btu)		7,161

All of the yields and utility requirements are similar between the model and the literature except cooling water. The uop_hdrd.bkp model predicts a much higher cooling water load than projected by UOP. This discrepancy can be due to many factors, including improved equipment design and heat integration in the UOP process and differing cooling water specifications (e.g., allowable temperature rise). The discrepancy was not explored further since cooling water is a very small contributor to the impacts in an LCA. Table 3 compares the results of this modeling effort and the values from the UOP report (2006). Carbon dioxide was not reported in the UOP study.

The propane mix gas is composed of 93.3% propane, 5.7% CO_2, and 1% water. The lower heating value (LHV) of the mix is estimated at 18,568 Btu/lb. The entire mass balance for the simulation is contained in Appendix A2-6.

In addition to the material and energy balance, the analysis projected the air emissions from the process. As noted earlier, it is assumed that the fuel gas, which is primarily propane, is combusted with make-up natural gas in order to meet the energy demand of the process. Thus, it was assumed that there were minimal air emissions from the main process. The LCA analysis will provide the emissions from the combustion of the fuel gas and any other fuel needed to generate the necessary steam and electricity. This assessment is outside the process lines for this process configuration.

The UOP HDRD process uses hydrogenation to convert bio-oils like soybean oil into a diesel substitute. Several companies are looking into this process. This analysis developed an ASPEN Plus® model of the process and compared its results with published results by UOP and NREL (UOP 2006). Good agreement was obtained between the two studies. These results will be used to develop an LCA for this process.

A2-5. References

Marker, T., 2006, Email to V. Putsche with cc to C. Johnson of NREL, "Follow-up on Green Diesel and Green Gasoline LCA Requests," Aug. 15.

Sheehan, J., V. Camobreco, J. Duffield, M. Graboski, H. Shapouri, 1998, *An Overview of Biodiesel and Petroleum Diesel Life Cycles*, joint study sponsored by U.S. Department of Agriculture and U.S. Department of Energy, NREL/TP-580-24772.

UOP, 2006, *Opportunities for Biorenewables in Oil Refineries*, Final Technical Report, DOE Award Number DE-FG36-05GO15085, contributors were Terry Marker, John Petri, Tom Kalnes, Micke McCall, Dave Mackowiak, Bob Jerosky, Bill Reagan, Lazlo Nemeth, Mark Krawczyk (UOP); Stefan Czernik (NREL); Doug Elliott (PNNL); David Shonnard (Michigan Technological University).

A2-6. ASPEN Plus® Input File: UOP_HDRD.inp

```
;
;Input Summary created by Aspen Plus Rel. 13.1 at 18:15:55 Fri Sep 22, 2006
;Directory  C:\AspenTech\Aspen  Plus  2004  Filename  C:\AspenTech\Aspen  Plus
2004\uop_hdrd.inp
;
TITLE 'HDRD - UOP'

IN-UNITS ENG DENSITY='lb/gal' POWER=kW VOLUME=gal & MOLE - DENSITY
=' lbmol/gal' MASS-DENSITY='lb/gal'
```

DEF-STREAMS CONVEN ALL

DATABANKS PURE13 / AQUEOUS / SOLIDS / INORGANIC / & NOASPENPCD

PROP-SOURCES PURE13 / AQUEOUS / SOLIDS / INORGANIC

COMPONENTS
H2 H2 /
LINOL3 C18H30O2 /
PALM C16H32O2 /
STEARIC C18H36O2 /
OLEIC C18H34O2 /
LINOL2 C18H32O2 /
ARACHID C20H40O2 /
GDSL C18H38 /
H2O H2O / H2S H2S / NH3 H3N / PROPANE C3H8 /
NAPTHA / CO2 CO2

PC-USER
IN-UNITS ENG
PC-DEF ASPEN GDSL GRAV=0.78 MW=254. PC-DEF ASPEN NAPTHA GRAV=0.7
MW=100.

ADA-SETUP
ADA-SETUP PROCEDURE=REL9

HENRY-COMPS HC-1 CO2

FLOWSHEET
BLOCK RX101 IN=105 119 OUT=106 QRX101
BLOCK S101 IN=108 OUT=115 110 109
BLOCK P101 IN=101 OUT=102 WP101
BLOCK P102 IN=112 OUT=113 19 WP102
BLOCK CP102 IN=117 OUT=118 WCP-102
BLOCK CP101 IN=120 OUT=121 WCP-101
BLOCK HX101+ IN=106 20 OUT=107 QHX101
BLOCK HX101- IN=103 QHX101 OUT=104 QHX101XS BLOCK HX102- IN=104
QHX102 OUT=105 QHX102XS BLOCK MX101 IN=102 OUT=103
BLOCK MX102 IN=118 121 OUT=119
BLOCK PSA IN=115 OUT=117 CO2 125
BLOCK ST-101 IN=114 110A OUT=111 112
BLOCK HX103+ IN=107 OUT=108 QHX103
BLOCK HX103- IN=CWS QHX103 OUT=CWR QHX103XS BLOCK HX102+
IN=MPSS OUT=MPSR QHX102
BLOCK SP-101 IN=110 OUT=110A 110B

BLOCK P105 IN=109 18 OUT=WWT WP-105 BLOCK P106 IN=110B OUT=110C
WP-106 BLOCK P103 IN=CWS1 OUT=CWS WP-103 BLOCK P104 IN=CWR
OUT=CWR1 WP-104 BLOCK FL-101 IN=125 111 OUT=PROPANE 17 16 BLOCK B8
IN=19 OUT=18
BLOCK B9 IN=QRX101 OUT=20 21

PROPERTIES POLYUF HENRY-COMPS=HC-1 PROPERTIES POLYNRTL

PROP-DATA HENRY-1
IN-UNITS ENG

PROP-LIST HENRY
BPVAL CO2 H2O 175.2762325 -15734.78987 -21.66900000 &
6.12550005E-4 31.73000375 175.7300026 0.0

STREAM 101
IN-UNITS ENG
SUBSTREAM MIXED TEMP=68. PRES=14.7 MASS-FLOW=100. MASS-FRAC
LINOL3 0.075 / PALM 0.11 / STEARIC 0.041 / & OLEIC 0.22 / LINOL2 0.54 /
ARACHID 0.014

STREAM 114
IN-UNITS ENG
SUBSTREAM MIXED TEMP=400. PRES=50. MASS-FLOW=2.8 MASS-FRAC H2O
1.

STREAM 117
IN-UNITS ENG
SUBSTREAM MIXED TEMP=68. PRES=14.7 MASS-FLOW=1. MASS-FRAC H2 1.

STREAM 120
IN-UNITS ENG
SUBSTREAM MIXED TEMP=68. PRES=14.7 MASS-FLOW=100. MASS-FRAC H2
1.

STREAM CWS
IN-UNITS ENG
SUBSTREAM MIXED TEMP=35. <C> PRES=500. MASS-FLOW=100. MASS-FRAC
H2O 1.

STREAM CWS1
IN-UNITS ENG
SUBSTREAM MIXED TEMP=35. <C> PRES=14.7 MASS-FLOW=100. MASS-FRAC
H2O 1.

STREAM MPSS
IN-UNITS ENG

SUBSTREAM MIXED TEMP=667. PRES=150. MASS-FLOW=100. MASS-FRAC
H2O 1.

DEF-STREAMS HEAT 20

DEF-STREAMS HEAT 21

DEF-STREAMS HEAT QHX101

DEF-STREAMS HEAT QHX101XS

DEF-STREAMS HEAT QHX102

DEF STREAMS HEAT QHX102XS

DEF-STREAMS HEAT QHX103

DEF-STREAMS HEAT QHX103XS

DEF-STREAMS HEAT QRX101

DEF-STREAMS WORK WCP-101

DEF-STREAMS WORK WCP-102

DEF-STREAMS WORK WP-103

DEF-STREAMS WORK WP-104

DEF-STREAMS WORK WP-105 DEF-STREAMS WORK WP-106 DEF-STREAMS
WORK WP101 DEF-STREAMS WORK WP102

BLOCK B8 MIXER

BLOCK MX101 MIXER IN-UNITS ENG

BLOCK MX102 MIXER IN-UNITS ENG

BLOCK B9 FSPLIT
FRAC 20 0.15

BLOCK SP-101 FSPLIT IN-UNITS ENG

FRAC 110A 0.99

BLOCK PSA SEP IN-UNITS ENG PARAM

FRAC STREAM=117 SUBSTREAM=MIXED COMPS=H2 H2O PROPANE CO2 &
FRACS=1. 0. 0. 0.
FRAC STREAM=CO2 SUBSTREAM=MIXED COMPS=PROPANE CO2 FRACS= &
0. 1.

BLOCK HX101+ HEATER
IN-UNITS ENG
PARAM TEMP=100. <C> PRES=500.

BLOCK HX101- HEATER
IN-UNITS ENG
PARAM TEMP=290. <C> PRES=500.

BLOCK HX102+ HEATER
IN-UNITS ENG
PARAM PRES=500. VFRAC=0.

BLOCK HX102- HEATER
IN-UNITS ENG
PARAM TEMP=325. <C> PRES=500.

BLOCK HX103+ HEATER
IN-UNITS ENG
PARAM TEMP=100. PRES=500.

BLOCK HX103- HEATER
IN-UNITS ENG
PARAM PRES=500. DELT=15.

BLOCK FL-101 FLASH2
PARAM TEMP=68. PRES=14.7

BLOCK-OPTION FREE-WATER=YES

BLOCK S101 FLASH2
IN-UNITS ENG
PARAM TEMP=100. PRES=175.

BLOCK-OPTION FREE-WATER=YES

BLOCK ST-101 RADFRAC

```
IN-UNITS ENG
PARAM NSTAGE=8
COL-CONFIG CONDENSER=NONE REBOILER=NONE FEEDS 114 9 / 110A 1
PRODUCTS 111 1 V / 112 8 L
P-SPEC 1 14.7
COL-SPECS DP-STAGE=1.


BLOCK RX101 RYIELD
IN-UNITS ENG
PARAM TEMP=325. <C> PRES=500.
MASS-YIELD MIXED GDSL 0.8415 / H2O 0.02125 / CO2 & 0.10625 / PROPANE
0.029 / H2 0.001


BLOCK P101 PUMP
IN-UNITS ENG
PARAM PRES=500.


BLOCK P102 PUMP
IN-UNITS ENG
PARAM DELP=10.


BLOCK-OPTION FREE-WATER=YES


BLOCK P103 PUMP
IN-UNITS ENG
PARAM PRES=500. PUMP-TYPE=TURBINE


BLOCK P104 PUMP
IN-UNITS ENG
PARAM DELP=10.


BLOCK P105 PUMP
IN-UNITS ENG
PARAM DELP=10.


BLOCK P106 PUMP
IN-UNITS ENG
PARAM DELP=10. PUMP-TYPE=PUMP


BLOCK CP101 COMPR
IN-UNITS ENG
PARAM TYPE=ISENTROPIC PRES=500. MODEL-TYPE=TURBINE


BLOCK CP102 COMPR
IN-UNITS ENG
PARAM TYPE=ISENTROPIC PRES=500. MODEL-TYPE=TURBINE
```

```
DESIGN-SPEC DS-HX101
IN-UNITS ENG
DEFINE        QXS        INFO-VAR        INFO=HEAT        VARIABLE=DUTY        &
STREAM=QHX101XS
SPEC "QXS" TO "0.0"
TOL-SPEC "0.1"
VARY BLOCK-VAR BLOCK=HX101+ VARIABLE=TEMP SENTENCE=PARAM
LIMITS "100" "617"


DESIGN-SPEC DS-HX102
IN-UNITS ENG
DEFINE STMIN STREAM-VAR STREAM=MPSS SUBSTREAM=MIXED &
VARIABLE=MASS-FLOW
DEFINE QXS INFO-VAR INFO=HEAT VARIABLE=DUTY &
STREAM = QHX102XS SPEC "QXS" TO "0"
TOL-SPEC "0.1"
VARY  STREAM-VAR  STREAM=MPSS  SUBSTREAM=MIXED  &  VARIABLE
=MASS-FLOW
LIMITS "0" "10000"


DESIGN-SPEC DS-HX103
IN-UNITS ENG
DEFINE CWIN STREAM-VAR STREAM=CWS1 SUBSTREAM=MIXED &
VARIABLE=MASS-FLOW
DEFINE QXS INFO-VAR INFO=HEAT VARIABLE=DUTY &
STREAM=QHX103XS SPEC "QXS" TO "0"
TOL-SPEC "0.1"
VARY        STREAM-VAR        STREAM=CWS1        SUBSTREAM=MIXED        &
VARIABLE=MASS-FLOW
LIMITS "100" "10000"


EO-CONV-OPTI


CALCULATOR H2IN
IN-UNITS ENG
DEFINE H2IN STREAM-VAR STREAM=120 SUBSTREAM=MIXED &
VARIABLE=MASS-FLOW
DEFINE OILIN STREAM-VAR STREAM=101 SUBSTREAM=MIXED &
VARIABLE=MASS-FLOW
F H2IN = 0.0272*OILIN
READ-VARS OILIN


CALCULATOR HYDCRK
IN-UNITS ENG
DEFINE FEED STREAM-VAR STREAM=101 SUBSTREAM=MIXED &
VARIABLE=MASS-FLOW
```

```
DEFINE    GDYLD    BLOCK-VAR    BLOCK=RX101    VARIABLE=YIELD    &
SENTENCE=MASS-YIELD ID1=MIXED ID2=GDSL
DEFINE    PROYLD    BLOCK-VAR    BLOCK=RX101    VARIABLE=YIELD    &
SENTENCE=MASS-YIELD ID1=MIXED ID2=PROPANE
DEFINE    CO2YLD    BLOCK-VAR    BLOCK=RX101    VARIABLE=YIELD    &
SENTENCE=MASS-YIELD ID1=MIXED ID2=CO2
DEFINE    H2OYLD    BLOCK-VAR    BLOCK=RX101    VARIABLE=YIELD    &
SENTENCE=MASS-YIELD ID1=MIXED ID2=H2O
DEFINE    FD105    STREAM-VAR    STREAM=105    SUBSTREAM=MIXED    &
VARIABLE=MASS-FLOW
DEFINE    FD119    STREAM-VAR    STREAM=119    SUBSTREAM=MIXED    &
VARIABLE=MASS-FLOW
DEFINE    H2IN    STREAM-VAR    STREAM=120    SUBSTREAM=MIXED    &
VARIABLE=MASS-FLOW
DEFINE    H2YLD    BLOCK-VAR    BLOCK=RX101    VARIABLE=YIELD    &
SENTENCE=MASS-YIELD ID1=MIXED ID2=H2
F TTLFD = FD105+FD119
F GDYLD = 0.828*(FEED+H2IN)/TTLFD
F PROYLD = 0.047*(FEED+H2IN)/TTLFD
F CO2YLD = 0.075*(FEED+H2IN)/TTLFD
F H2OYLD = 0.050*(FEED+H2IN)/TTLFD
F SUM = GDYLD+PROYLD+CO2YLD+H2OYLD
F DIFF = TTLFD - (SUM*TTLFD)
F H2YLD = DIFF/TTLFD
F WRITE(NHSTRY,*)SUM,DIFF,H2YLD
READ-VARS FEED FD105 FD119 H2IN
WRITE-VARS GDYLD PROYLD CO2YLD H2OYLD H2YLD BLOCK-OPTION
SIM-LEVEL=4

TEAR

TEAR 117
STREAM-REPOR NOMOLEFLOW MASSFLOW
```

PROPERTY-REP NOPARAM-PLUS

	101	102	103	104	105	106	107	108	109	110
	P101	MX101	HX101-	HX102-	RX101	4□	;□	S101	P105	SP-101
		P101	MX101	HX101-	HX102-	RX101	4□	4□	S101	S101
	LIQUID	LIQUID	LIQUID	LIQUID	LIQUID	MIXED	MIXED	MIXED	LIQUID	LIQUID
Substream: MIXED										
Mass Flow lb/hr										
H2	0	0	0	0	0	1.084999	1.084999	1.084999	0	2.04E-07
LINOL3	7.5	7.5	7.5	7.5	7.5	0	0	0	0	0
PALM	11	11	11	11	11	0	0	0	0	0
STEARIC	4.1	4.1	4.1	4.1	4.1	0	0	0	0	0
OLEIC	22	22	22	22	22	0	0	0	0	0
LINOL2	54	54	54	54	54	0	0	0	0	0
ARACHID	1.4	1.4	1.4	1.4	1.4	0	0	0	0	0
GDSL	0	0	0	0	0	85.05216	85.0216	85.05216	0	85.052
H2O	0	0	0	0	0	5.54688	5.54688	5.54688	5.445815	0.0218078
H2S	0	0	0	0	0	0	0	0	0	0
NH3	0	0	0	0	0	0	0	0	0	0
PROPANE	0	0	0	0	0	4.82784	4.82784	4.82784	0	2.468692
NAPTHA	0	0	0	0	0	0	0	0	0	0
CO2	0	0	0	0	0	7.29312	7.29312	7.29312	0	0.2876933
Total Flow lbmol/hr	0.3591571	0.3591571	0.3591571	0.3591571	0.3591571	1.456175	1.456175	1.456175	0.3022887	0.3985819
Total Flow lb/hr	100	100	100	100	100	103.805	103.805	103.805	5.445815	87.8302
Total Flow cuft/hr	1.786671	1.795957	1.795957	1.975812	2.061152	29.28627	25.1053	10.49892	0.0877934	1.907063
Temperature F	68	80.58947	80.58948	290	370	617	507.6096	100	100	100
Pressure psi	14.7	500	500	500	500	500	500	500	175	175
Vapor Frac	0	0	0	0	0	0.8408124	0.7704222	0.4824463	0	0
Liquid Frac	1	1	1	1	1	0.1591876	0.2295778	0.5175537	1	1

PROPERTY-REP NOPARAM-PLUS (Cont.)

	101	102	103	104	105	106	107	108	109	110
	P101	MX101	HX101-	HX102-	RX101	4□	;□	S101	P105	SP-101
		P101	MX101	HX101-	HX102-	RX101	4□	4□	S101	S101
	LIQUID	LIQUID	LIQUID	LIQUID	LIQUID	MIXED	MIXED	MIXED	LIQUID	LIQUID
Solid Frac	0	0	0	0	0	0	0	0	0	0
Enthalpy Btu/lbmol	-3.14E+05	-3.12E+05	-3.12E+05	-2.83E+05	-2.71E+05	-70801.0⁻	-77885.8	-1.00E+05	-1.22E+05	-1.99E+05
Enthalpy Btu/lb	-1126.343	-1120.916	-1120.916	-1017.751	-971.6202	-993.196	-1092.581	-1406.734	-6797.624	-903.7778
Enthalpy Btu/hr	-1.13E+05	-1.12E+05	-1.12E+05	-1.02E+05	-97162.02	-1.03E+05	-1.13E+05	-1.46E+05	-37018.6	-79378.98
Entropy Btu/lbmol-R	-415.036	-412.2374	-412.2374	-367.7857	-351.4557	-80.6600	-87.58823	-118.2502	-38.21506	-382.6785
Entropy Btu/lb-R	-1.490631	-1.48058	-1.48058	-1.320929	-1.262278	-1.13149⁻	-1.228686	-1.658812	-2.121258	-1.736632
Density lbmol/gal	0.0268725	0.0267335	0.0267335	0.0243	0.0232939	6.65E-03	7.75E-03	0.0185411	0.4602862	0.0279396
Density lb/gal	7.482102	7.443416	7.443416	6.765855	6.485721	0.473829₃	0.5527402	1.321727	8.292184	6.156685
Average MW	278.4297	278.4297	278.4297	278.4297	278.4297	71.2860₉	71.28609	71.28609	18.01528	220.3567
Liq Vol 60F cuft/hr	1.803603	1.803603	1.803603	1.803603	1.803603	2.596039₃	2.596039	2.596039	0.087017	1.834368
	110A	110B	110C	111	112	113	114	115	117	118
	ST-101	P106	MX101	FL-101	P102		ST-101	PSA	CP102	MX102
	SP-101	SP-101	P106	ST-101	ST-101	P102		S101	PSA	CP102
	LIQUID	MISSING	MISSING	VAPOR	LIQUID	LIQUID	VAPOR	VAPOR	VAPOR	VAPOR

Substream: MIXED

Mass Flow lb/hr

	101	102	103	104	105	106	107	108	109	110
H2	2.04E-07	0	0	2.04E-07	4.05E-35	4.05E-35	0	1.08498	1.084999	1.084999
LINOL3	0	0	0	0	0	0	0	0	0	0
PALM	0	0	0	0	0	0	0	0	0	0
STEARIC	0	0	0	0	0	0	0	0	0	0
OLEIC	0	0	0	0	0	0	0	0	0	0
LINOL2	0	0	0	0	0	0	0	0	0	0

PROPERTY-REP NOPARAM-PLUS (Cont.)

	101	102	103	104	105	106	107	108	109	110
	P101	MX101	HX101-	HX102-	RX101	4□	;□	S101	P105	SP-101
		P101	MX101	HX101-	HX102-	RX101	4□	4□	S101	S101
	LIQUID	LIQUID	LIQUID	LIQUID	LIQUID	MIXED	MIXED	MIXED	LIQUID	LIQUID
ARACHID	0	0	0	0	0	0	0	0	0	0
GDSL	85.052	0	0	1.60E-04	85.05184	85.05184	0	1.58E-04	0	0
H2O	0.0218078	0	0	0.4588025	2.363005	0.029885	2.8	0.0792568	0	0
H2S	0	0	0	0	0	0	0	0	0	0
NH3	0	0	0	0	0	0	0	0	0	0
PROPANE	2.468692	0	0	2.321827	0.1468655	0.1468655	0	2.35948	0	0
NAPTHA	0	0	0	0	0	0	0	0	0	0
CO2	0.2876933	0	0	0.2876933	4.45E-12	4.45E-12	0	7.005427	0	0
Total Flow lbmol/hr	0.3985819	0	0	0.0846584	0.469347	0.3398392	0.1554236	0.7553041	0.5382258	0.5382258
Total Flow lb/hr	87.8302	0	0	3.068483	87.56171	85.22859	2.8	10.52899	1.084999	1.084999
Total Flow cuft/hr	1.907065	0	0	34.10666	1.85258	1.832826	19.0754	25.87814	18.60182	8.231991
Temperature F	100.002			98.90795	143.0657	149.9646	297.7949	100	99.97435	240.933
Pressure psi	175		500	14.7	21.7	31.7	64.7	175	175	500
Vapor Frac	0	0		1	0	0	1	1	1	1
Liquid Frac	1	0		0	1	1	0	1	0	0
Solid Frac	0	0		0	0	0	0	0	0	0
Enthalpy Btu/lbmol	-1.99E+05			-72054.6	-1.90E+05	-2.16E-05	-1.02E+05	-39284.66	163.5656	1154.758
Enthalpy Btu/lb	-903.7778			-1987.963	-1018.401	-861.4131	-5676.514	-2818.112	81.13857	572.831
Enthalpy Btu/hr	-79378.98			-6100.031	-89172.93	-73417.02	-15894.24	-2967186	88.03524	621.5208
Entropy Btu/lbmol-R	-382.6781			-40.91684	-319.5699	-425.8229	-10.82218	-7.527309	-4.635141	-5.164754
Entropy Btu/lb-R	-1.73663			-1.128883	-1.712954	-1.69792	-0.6007222	-0.5399766	-2.299314	-2.562034
Density lbmol/gal	0.0279396			3.32E-04	0.0338676	0.0247868	1.09E-03	3.90E-03	3.87E-03	8.74E-03
Density lb/gal	6.156679			0.0120268	6.318376	6.216305	0.0196224	0.0543903	7.80E-03	0.0176194

	101	102	103	104	105	106	107	108	109	110
	P101	MX101	HX101-	HX102-	RX101	4□	;□	S101	P105	SP-101
		P101	MX101	HX101-	HX102-	RX101	4□	4□	S101	S101
	LIQUID	LIQUID	LIQUID	LIQUID	LIQUID	MIXED	MIXED	MIXED	LIQUID	LIQUID
Average MW	220.3567			36.24543	186.5607	250.7909	18.01528	13.94006	2.01588	2.01588
Liq Vol 60F cuft/hr	1.834368	0	0	0.0864746	1.792832	1.755387	0.0449381	0.6742691	0.4617512	0.4617512
	119	120	121	122	123	124	125	CO2	CWR	CWR1
	RX101	CP101	MX102	FL-101	FL-101	P105	FL-101	PSA	CWR	
	MX102		CP101			P102	PSA		P104	P104
									HX103-	
	VAPOR	VAPOR	VAPOR	LIQUID	LIQUID	LIQUID	MIXED	VAPOR	LIQUID	LIQUID

Substream: MIXED
Mass Flow lb/hr

	101	102	103	104	105	106	107	108	109	110
H2	3.804999	2.72	2.72	0	0	0	0	0	0	0
LINOL3	0	0	0	0	0	0	0	0	0	0
PALM	0	0	0	0	0	0	0	0	0	0
STEARIC	0	0	0	0	0	0	0	0	0	0
OLEIC	0	0	0	0	0	0	0	0	0	0
LINOL2	0	0	0	0	0	0	0	0	0	0
ARACHID	0	0	0	0	0	0	0	0	0	0
GDSL	0	0	0	2.84E-04	0	0	1.58E-04	0	0	0
H2O	0	0	0	4.63E-08	0.4894927	2.33312	0.0792568	0	2309.68	2309.68
H2S	0	0	0	0	0	0	0	0	0	0
NH3	0	0	0	0	0	0	0	0	0	0
PROPANE	0	0	0	1.48E-05	0	0	2.359148	0	0	0
NAPTHA	0	0	0	0	0	0	0	0	0	0

PROPERTY-REP NOPARAM-PLUS (Cont.)

	101	102	103	104	105	106	107	108	109	110
	P101	MX101 P101	HX101- MX101	HX102- HX101-	RX101 HX102-	4□ RX101	;□ 4□	S101 4□	P105 S101	SP-101 S101
	LIQUID	LIQUID	LIQUID	LIQUID	LIQUID	MIXED	MIXED	MIXED	LIQUID	LIQUID
CO_2	0	0	0	3.64E-08	0	0	0	7.005427	0	0
Total Flow lbmol/hr	1.887512	1.349287	1.349287	1.46E-06	0.0271709	0.1295079	0.0578996	0.1591788	128.2067	128.2067
Total Flow lb/hr	3.804999	2.72	2.72	2.99E-04	0.4894927	2.33312	2.438563	7.005427	2309.68	2309.68
Total Flow cuft/hr	42.90768	520.0834	34.66854	6.53E-06	7.85E-03	0.0381251	1.507645	5.167461	37.93036	37.93266
Temperature F	586.7733	68	724.6138	68	68	149.9646	99.97435	99.97435	110.4626	110.571
Pressure psi	500	14.7	500	14.7	14.7	31.7	175	175	500	510
Vapor Frac	1	1	1	0	0	0	0.9295689	1	0	0
Liquid Frac	0	0	0	1	1	1	0.070431	0	1	1
Solid Frac	0	0	0	0	0	0	0	0	0	0
Enthalpy Btu/lbmol	3577.495	-61.40031	4543.916	-1.88E+05	-1.23E+05	-1.22E+05	-50942.12	-1.69E+05	-1.22E+05	-1.22E+05
Enthalpy Btu/lb	1774.657	-30.45832	2254.061	-914.9968	-6829.944	-748.178	-1209.537	-3843.429	-6786.474	-6786.371
Enthalpy Btu/hr	6752.566	-82.84663	6131.045	-0.2738398	-3343.207	-15744.31	-2949.533	-2692486	-1.57E+07	-1.57E+07
Entropy Btu/lbmol-R	-2.354373	0.1171418	-1.486812	-363.1769	-39.26957	-36.67663	-66.63749	-4.062922	-37.84993	-37.8467
Entropy Btu/lb-R	-1.167913	0.0581095	-0.7375498	-1.771269	-2.179792	-2.035862	-1.582198	-0.0923185	-2.10099	-2.100811
Density lbmol/gal	5.88E-03	3.47E-04	5.20E-03	0.02986	0.4624672	0.4541016	5.13E-03	4.12E-03	0.4518477	0.4518203
Density lb/gal	0.0118546	6.99E-04	0.0104882	6.122437	8.331475	8.180767	0.2162236	0.1812281	8.140163	8.139668
Average MW	2.01588	2.01588	2.01588	205.0377	18.01528	18.01528	42.11703	44.0098	18.01528	18.01528
Liq Vol 60F cuft/hr	1.619323	1.157571	1.157571	6.32E-06	7.86E-03	0.037445	0.0759562	0.1365617	37.06883	37.06883

		CWS HX103-P103 LIQUID	CWS1 P103 LIQUID	MPSR 4□ LIQUID	MPSS 4□ VAPOR	PROPANE FL-101 VAPOR	WWT P105 LIQUID
Substream: MIXED							
Mass Flow	lb/hr						
H2		0	0	0	0	2.04E-07	0
LINOL3		0	0	0	0	0	0
PALM		0	0	0	0	0	0
STEARIC		0	0	0	0	0	0
OLEIC		0	0	0	0	0	0
LINOL2		0	0	0	0	0	0
ARACHID		0	0	0	0	0	0
GDSL		0	0	0	0	3.35E-05	0
H2O		2309.68	2309.68	5.372242	5.372242	0.0485666	7.778936
H2S		0	0	0	0	0	0
NH3		0	0	0	0	0	0
PROPANE		0	0	0	0	4.68096	0
NAPTHA		0	0	0	0	0	0
CO2		0	0	0	0	0.2876933	0
Total Flow lbmol/hr		128.2067	128.2067	0.2982047	0.2982047	0.1153857	0.4317965
Total Flow lb/hr		2309.68	2309.68	5.372242	5.372242	5.017253	7.778936
Total Flow cuft/hr		37.61554	37.60596	0.1048887	15.27979	43.78178	0.1280346
Temperature F		95.46259	95	366.04	366.0404	68	114.4461
Pressure	psi	500	14.7	164.7	164.7	14.7	41.7
Vapor Frac		0	0	0	1	1	0
Liquid Frac		1	1	1	0	0	1
Solid Frac		0	0	0	0	0	0
Enthalpy	Btu/lbmol	-1.23E+05	-1.23E+05	-1.17E+05	-1.02E+05	-53610.59	-1.22E+05
Enthalpy	Btu/lb	-6800.593	-6801.025	-6510.467	-5651.786	-1232.925	-6782.69
Enthalpy	Btu/hr	-1.57E+07	-1.57E+07	-34975.8	-30362.76	-6185.896	-52762.11
Entropy	Btu/lbmol-R	-38.29923	-38.31317	-30.87826	-12.05098	-59.06788	-37.73145
Entropy	Btu/lb-R	-2.125931	-2.126704	-1.714004	-0.6689311	-1.35843	-2.094414
Density	lbmol/gal	0.4556294	0.4557454	0.3800617	2.61E-03	3.52E-04	0.4508374
Density	lb/gal	8.208291	8.210381	6.846918	0.0470009	0.0153193	8.121961
Average MW		18.01528	18.01528	18.01528	18.01528	43.48245	18.01528
Liq Vol 60F cuft/hr		37.06883	37.06883	0.0862209	0.0862209	0.1545686	0.1248467

In: Biodiesel Fuels Reexamined
Editor: Bryce A. Kohler

ISBN: 978-1-60876-140-1
© 2011 Nova Science Publishers, Inc.

Chapter 4

EMPIRICAL STUDY OF THE STABILITY OF BIODIESEL AND BIODIESEL BLENDS: MILESTONE REPORT[*]

R.L. McCormick and S.R. Westbrook

ACKNOWLEDGMENTS

This study is the result of a collaborative program between the National Renewable Energy Laboratory (NREL) and the National Biodiesel Board (NBB), with technical direction from the ASTM Biodiesel Stability Task Force. The authors acknowledge financial support from NBB and technical assistance from NBB technical director Steve Howell. Additionally the authors thank Teresa Alleman, Stu Porter, and Melissa Williams for their assistance in completing this project.

EXECUTIVE SUMMARY

In support of the U.S. Department of Energy Fuels Technologies Program Multiyear Program Plan goal of identifying fuels that can displace 5% of petroleum diesel by 2010, the National Renewable Energy Laboratory (NREL), in collaboration with the National Biodiesel Board (NBB) and with subcontractor Southwest Research Institute, performed a study of biodiesel oxidation stability. The objective of this work was to develop a database that supports specific proposals for a stability test and specification for biodiesel and biodiesel blends. B100 samples from 19 biodiesel producers were obtained in December of 2005 and January of 2006 and tested for stability. Eight of these samples were then selected for additional study, including long-term storage tests and blending at 5% and 20% with a number of ultra-low sulfur diesel (ULSD) fuels. These blends were also tested for stability.

---

[*] This is an edited, reformatted and augmented version of a national laboratory of the U.S. Department of Energy's publication, dated May 2007.

The study used accelerated tests as well as tests that were intended to simulate three real-world aging scenarios: (1) storage and handling, (2) vehicle fuel tank, and (3) high-temperature engine fuel system. Several tests were also performed with two commercial antioxidant additives to determine whether these additives improve stability. This report documents completion of NREL's Fiscal Year 2007 Annual Operating Plan Milestone 10.1.

The B100 samples examined show a broad distribution of stability on accelerated tests, with oil stability index (OSI) or Rancimat induction time results ranging from less than 1 hour to more than 9 hours and ASTM D2274 total insolubles ranging from less than 2 mg/100 ml to nearly 18 mg/100 ml. The accelerated test data indicate that if the B100 stability is above roughly a 3-hour induction time, blends prepared from that B100 appear to be stable on the OSI and D2274 tests.

The D4625 long-term storage results for B100 indicate that most biodiesel samples, regardless of initial induction time, will begin to oxidize immediately during storage. If induction time is near or below the 3-hour limit, the B100 will most likely go out of specification for either stability or acid value within 4 months. Even B100 with induction times longer than 7 hours will be out of specification for oxidation stability at only 4 months, although these samples may not have shown a significant increase in acidity or in deposit formation. The 3-hour B100 induction time limit appears to be adequate to prevent oxidative degradation for both B5 and B20 blends in storage for up to 12 months.

For tests that simulated fuel tank aging and high temperature stability, we conclude that stable B100 (longer than 3 hours induction time) leads to stable B5 blends. For B20, the results are less definitive, but provide considerable evidence that B100 with induction time of at least 3 hours produces stable B20 blends, but the test cannot differentiate between intermediate and highly stable samples for acid number increase or sediment formation under these worst-case test conditions. Additional work is required to confirm this finding and to determine whether an additional stability test for the B20 blend is required.

These results indicate that B100 stability is the main factor that affects the stability of B5 and B20 blends, independent of diesel fuel aromatic content, sulfur level, or stability. An antagonism between unstable B100 and diesel fuel was observed. For B100 with an induction time lower than 3-hours, the level of deposits formed on the D2274 stability test was well above what would be expected based on the B100 deposit level and the percentage of biodiesel in the fuel. This antagonistic effect was significantly greater for B20 than for B5. The hindered phenolic antioxidants tested here prevented oxidative degradation of B100 in the storage simulation, and prevented degradation of biodiesel blends in the storage, fuel tank, and high-temperature simulations.

We recommend that additional tests be performed with real equipment to validate these conclusions.

ACRONYMS AND ABBREVIATIONS

ASTM	ASTM International, a standards setting organization
Bxx	designation of biodiesel or biodiesel blend (B100 is pure biodiesel, B20 is a 20% blend, for example)
EN	European Normalisation

ISO	International Standards Organization
KOH	potassium hydroxide
NBB	National Biodiesel Board
NREL	National Renewable Energy Laboratory
OSI	oil stability index (also called Rancimat induction time)
PV	peroxide value
RME	rapeseed methyl ester
T90	90% boiling temperature
ULSD	ultra-low sulfur diesel

INTRODUCTION

In support of the U.S. Department of Energy Fuels Technologies Program Multiyear Program Plan Goal of identifying fuels that can displace 5% of petroleum diesel by 2010, the National Renewable Energy Laboratory (NREL), in collaboration with the National Biodiesel Board (NBB) and with subcontractor Southwest Research Institute, performed a study of biodiesel oxidation stability. The objective of this work was to develop a database to support specific proposals for a stability test and specification for biodiesel and biodiesel blends. B100 samples from 19 biodiesel producers were obtained during December 2005 and January 2006 and tested for stability. Eight of these samples were then selected for additional study, including long-term storage tests and blending at 5% and 20% with a number of ultra-low sulfur diesel (ULSD) fuels. These blends were also tested for stability. The study employed accelerated tests as well as tests intended to simulate three real-world aging scenarios: (1) storage and handling, (2) vehicle fuel tank, and (3) high-temperature engine fuel system. Results were analyzed to determine whether ensuring B100 stability was adequate to ensure the stability of B5 and B20 blends. Several tests were also performed with two commercial antioxidant additives to determine whether these additives might improve stability. This report documents completion of the NREL Fiscal Year 2007 Annual Operating Plan Milestone 10.1.

BACKGROUND

Biodiesel produced from vegetable oils and other feedstocks can be more prone to oxidation than a typical petroleum diesel unless it is modified or treated with additives. The purpose of this project is to provide data that define how stable a biodiesel or biodiesel blend should be to prevent the formation of acids or sediment during transportation, storage, and use; and subsequent vehicle operational and maintenance problems.

The general mechanism of fat (or lipid) oxidation is reasonably well understood.[1,2] The fatty acid alkyl chains have varying numbers of double bonds. When multiple double bonds are present they are in allylic configuration, which means that they are separated by a single methylene group called a *bis*-allylic carbon or position. For biodiesel made from common feedstocks, such as oils from soy, rapeseed, and palm, as well as lard and tallow, the fatty acid chains contain primarily 16 or 18 carbon atoms and from zero to three double bonds.

Eighteen-carbon chains contain one double bond for oleic acid, two for linoleic acid, and three for linolenic acid. The relative oxidation rates for these C18 esters are linolenic > linoleic >> oleic[3] because the di- and tri- unsaturated fatty acids contain the most reactive sites for initiating the auto-oxidation chain reaction sequence. Oxidation rate correlates with the total number of bis-allylic sites (the methylene CH directly adjacent to the two double bonds), not with the total number of double bonds.[4] These sites have the highest reactivity for the for "formation"mation of free radicals that can react directly with oxygen to form peroxide radicals.

This is the initial step in the classic auto-oxidation mechanism with the chain reaction steps of initiation, propagation, chain branching, and termination.[1,2] The allylic position (a methylene adjacent to a single double bond) is much less reactive, explaining the much lower oxidation rate of oleic acid. Also, the radicals formed at the bis-allylic sites immediately isomerize to form a more stable conjugated structure. The peroxide radicals can cleave to form acids and aldehydes, or can react with another fatty acid chain to form a dimer. Peroxide can form via this route even at ambient temperature.

Several studies have examined biodiesel stability. Westbrook[5] used the ASTM D4625 test wherein the subject fuel or blending component is held for 12 weeks at 43°C in a capped but vented glass bottle, and oxidation effects (the amount of sediment formed, acidity, and change in viscosity) are periodically measured. Wide variations of insolubles formation, acid number, and viscosity increase were observed; the least stable samples exhibited unacceptable levels of insolubles and acidity 4 to 8 weeks into the test. To more fully understand the causes of these differences, McCormick and coworkers[6] examined the stability characteristics of biodiesel samples that were commercially available at blenders and distributors during 2004 and showed that the stability range results primarily from differences in fatty acid makeup and natural antioxidant content. However, samples containing high out-of-specification levels of glycerides (unconverted or partly converted feedstock) also tend to form deposits. Storage conditions can have a large impact on the degree of oxidation. Thus, for real-world samples containing impurities, the correlation of oxidation rate or tendency with the number of bis-allylic sites may be skewed or overshadowed by other factors such as natural antioxidant content.

Bondioli and coworkers[7] stored several drum quantity samples of biodiesel at ambient conditions for 1 year. One sample was "shaken" once per week to promote intimate contact with air. Over this period the quiescent samples exhibited little or no change in properties, including only minor reductions in Rancimat induction time. This contrasts strongly with results of studies conducted at higher temperatures (for example 43°C as in ASTM D4625), which have shown large changes in acid value and other parameters.[5,8] Additionally, the agitated sample exhibited significant increase in peroxide and acid values and a large reduction in Rancimat induction time because of the increased exposure to dissolved oxygen. Mittelbach and Gangl also stored biodiesel produced from rapeseed oil and used frying oil under different conditions for up to 200 days.[9] Degradation caused by oxidation began immediately, as shown by the formation of peroxides and reduction in Rancimat induction time. However, even at the end of the storage period, limits for viscosity and acidity were not exceeded.

Fang and McCormick[10] showed that dimerization of the peroxide species is not the only mechanism for molecular weight growth and deposit formation in biodiesel, and identified several other mechanisms by which biodiesel can degrade. In particular, aldehydes formed by peroxide decomposition could also polymerize via aldol condensation. However, all pathways to deposit formation involved peroxide formation as the initial step, highlighting the importance of preventing peroxide formation at the point of biodiesel manufacture and throughout the biodiesel distribution chain. Fang and McCormick also showed an antagonistic or synergistic effect on deposit formation for biodiesel blends. The formation of deposits under oxidative stressing was higher than predicted for biodiesel blends based on the weight percent biodiesel and the deposits formed from the diesel fuel or biodiesel alone. This effect was most significant for blends in the 20% to 30% range.

Vegetable oils contain naturally occurring antioxidants that can cause oxidation stability to vary over a wide range, even for samples with similar fatty acid makeup.[6] The most common antioxidants are tocopherols. For example, soy oils contain 500 to 3000 ppm tocopherols, along with other antioxidants such as sterols and tocotrienols,[11] which may not be affected by the ester preparation process.[12] However, some production processes include steps to purify the methyl esters that can remove antioxidants. Biodiesel produced using distillation to purify the product, for example, typically contains little or no natural antioxidant and is less stable than biodiesel that does. The stability of biodiesel prepared by distillation, as well as biodiesel that contains natural antioxidants, can be improved by adding synthetic antioxidants.[2] Westbrook[5] has shown that several antioxidants, including proprietary antioxidants developed for petroleum fuels, can improve biodiesel oxidation stability.

High levels of oxidation can cause operational problems for engine fuel system components. Terry and coworkers[13] showed that at very high levels of oxidation, biodiesel blends can separate into two phases to cause fuel pump and injector operational problems or lacquer deposits on fuel system components. Blassnegger performed 500-hour fuel injector bench tests with rapeseed methyl ester (RME) (B100) samples that had a range of stability.[14] Lower stability B100, with a Rancimat induction time of 1.8 to 3.5 hours, produced injector deposits and reduced the amount of fuel injected per stroke. Tsuchiya et al demonstrated that the formation of high levels of acids, which can be generated from unstable biodiesel during extended fuel recirculation, can corrode metal vehicle fuel tanks.[15] Thus, there are strong arguments for ensuring the stability of biodiesel and biodiesel blends. We conducted this empirical study, which is directed at defining a performance-based standard, because of the numerous factors can affect oxidation

METHODS

B100 and Diesel Fuel Properties

During December of 2005 and January of 2006 a single drum of B100 was obtained from each of 14 biodiesel production sites in the United States and two more from Canadian production facilities. Additionally, three drums of European rapeseed-derived biodiesel (RME) were obtained. The 19 B100 drums were nitrogen purged, and blanketed, and stored in a dark room at room temperature. All B100 samples were tested initially for:

- Total acid number (or acid value), ASTM D664
- Free and total glycerin, ASTM D6584.

Detailed characterization was performed on a subset of eight fuels that were selected for more detailed testing:

- Total particulate contamination, ASTM D6217
- Flash point, ASTM D93
- Peroxide value, ASTM D3703
- Karl Fischer moisture, ASTM D6304
- Iso-octane insoluble, (based on a 4:1 dilution of the sample with iso-octane followed by filtration using the modified ASTM D2274 procedures described below)
- Kinematic viscosity @ 40°C, ASTM D 445
- Polymer content, ISO 16931 (high performance size-exclusion chromatography)
- Metals (P, Na, K, Ca, Mg, Cu, Zn), ASTM D5185.

In addition, six samples (two drums each) of petroleum-derived diesel fuel were obtained from petroleum refiners in the United States and Canada. These include one 500-ppm sulfur fuel, and five others meeting the 15-ppm sulfur limit (ULSD). The diesel fuels were characterized with the following tests:

- Total particulate contamination, ASTM D6217
- Flash point, ASTM D93
- Sulfur, ASTM D5453
- T90/carbon residue, ASTM D86/D524
- Total acid number (or acid value), ASTM D664
- Peroxide value, ASTM D3703
- Ash content, ASTM D482
- Supercritical fluid chromatography aromatics, ASTM D5186.

Peroxide is measured by D3703, a method developed for aviation fuels with peroxide contents in the range of 1 to 5 ppm. The D3703 method is therefore not wholly applicable to samples with the high peroxide numbers found in many B100 and biodiesel blend samples. An interlaboratory study is underway to determine the applicability of this method to biodiesel. The absolute accuracy of the test for biodiesel is currently unknown, but we believe it is adequate for showing trends over time.

Accelerated Stability Tests

All samples, B100 and blends, were analyzed under the same conditions per standard test method EN 14112 Fat and oil derivatives—Fatty acid methyl esters (FAME)—Determination of Oxidation Stability. The Rancimat apparatus was used and tests were performed at 110°C. The Rancimat utilizes a proprietary computer algorithm to analyze the conductivity results and determine the induction period. All samples were tested in duplicate and the average induction period was reported. If the results for the replicates differed by more than approximately 10% of the mean, a third test was conducted and the two results closest to each other were averaged.

Significant problems were often encountered when analyzing blends, especially B5 blends. For most blends, the petroleum diesel fraction was sufficiently volatile to evaporate and condense in either the air tubes or the measuring vessel. Some of the plastic components of the Rancimat were incompatible with the petroleum diesel. This caused swelling of the parts and resultant air leaks. We had to replace these parts on a frequent basis. The standard air tube would soften and restrict the flow of air to the measuring vessel. We found that standard TygonTM laboratory tubing was an acceptable replacement for the original air tubing. As of this writing, the manufacturer is evaluating replacement parts that are compatible with petroleum for the analysis of blends.

Also, the test results for the blends were found to be far less repeatable. In many cases, the B5 samples would run for several days before an induction period was determined. This was due to the lower amount of biodiesel in the sample (compared to B100) and varied with the stability of the original B100. For the purposes of this project, we manually stopped Rancimat tests once the induction period was found to be greater than 12 hours.

Oxidation stability as determined by ASTM D2274 was modified for use with biodiesel by substituting glass fiber filters for the cellulose ester filters used with petroleum fuels, as described by Stavinoha and Howell.[16] Additionally, we attempted to keep the filtration time as consistent as possible. Iso-octane insolubles of the filtrate were determined by adding 100 ml of the filtered, post-D2274 B100 to 400 ml of prefiltered iso-octane. This was thoroughly mixed and allowed to set for at least 1 hour before filtering to determine insolubles. These insolubles or precipitates are a potential measure of oxidation products that remain soluble in the very polar B100, but could be insoluble in a diesel fuel blend. Acid value of the filtrate was also measured by D664 for some samples.

Thermal stability as determined by ASTM D6468 was performed at 150°C for 180 minutes. In addition to measurement of filter reflectance, this test was modified to use the same gravimetric insoluble measurement procedures used for D2274 (both total and iso-octane). Duplicate 50-ml samples were tested; results for reflectance were reported as the average. Gravimetric results are reported as the sum of the duplicate samples in mg/100 ml. An additional modification to D6468 was examined in preliminary scoping tests to purge the ullage or space above the heated liquid with air to ensure that a highly unstable material would not consume all the oxygen and thereby produce an invalid result. Results of these scoping tests are shown in Table 1, which shows that ullage purge had no significant effect on the results, and thus was not used in subsequent tests.

Table 1. Results of D6468 Ullage Purge Scoping Tests

Sample Identification	AL-27131-F[(1)]		AL-27070-F[(2)]	
Modified ASTM D6468[(3)]	with air flow	without air flow	with air flow	without air flow
filterable insolubles, mg/100 ml	0.03	0.14	0.77	1.1
adherent insolubles, mg/100 ml	0.09	0.06	0.14	0.14
total insolubles, mg/100 ml	0.11	0.2	0.9	1.2

[1] Soy-based B100
[2] Low-sulfur No. 2 diesel fuel
[3] 350 ml; 150°C; 180 minutes with and without air flow within ullage

Oxidation stability as determined by ASTM D525 Standard Test Method for Oxidation Stability of Gasoline (Induction Period Method), measures an induction time for the start of oxygen consumption. This test was shown in a recent study[17] to correlate well with Rancimat induction time.

Simulation of Real-World Aging Scenarios

We envision three scenarios where oxidative degradation of biodiesel or biodiesel blends might occur.

Storage and Handling
Biodiesel and biodiesel blends must be stored and handled before they are dispensed into vehicle fuel tanks. For research purposes there seems to be general agreement that ASTM D4625 (43°C/12 weeks) adequately simulates this situation.* The method employs the glass vessel shown in Figure 1. Individual glass vessels are stored for 4, 8, and 12 weeks so that an independent sample is evaluated at each time interval (as opposed to removing an aliquot from one vessel at each time interval). In this study the procedures were modified to employ glass fiber filters to determine insolubles. For biodiesel and blends, acid value, peroxide value, polymer content, and viscosity, as well as total insolubles, were measured during this test. For B100, iso-octane insolubles were also measured.

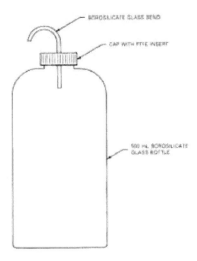

Figure 1. Glass vessel used for D4625 stability test.

The goals for this test are to:

- Determine whether there is an accelerated test that predicts D4625 results.
- Assess the potential of a B100 accelerated stability measurement to predict blend stability in this scenario.
- Reveal how antioxidants can affect storage stability.

Vehicle Fuel Tank

Biodiesel blends will ultimately be pumped into a vehicle fuel tank and will be subject to periodic heating to higher than ambient temperatures. This can be due to recirculation of hot fuel back into the fuel tank, or simply exposure to hot climates. The ASTM Biodiesel Stability Working Group discussed the possibility of storing the fuel for about 6 days at 80 C to simulate a type of worst-case situation. Although there is no ASTM test with this time and temperature duration, the ASTM D4625 procedures and apparatus to can be modified slightly to determine total insolubles at the end of this period. The measurement of acid value, peroxide value, polymer content, and viscosity may also prove valuable. Because the biodiesel is already blended with diesel fuel, the measurement of iso-octane insolubles is not likely to be informative. This is not a standard test, so results should be interpreted with caution.

A scoping study was performed to ensure that this test is not being conducted under oxygen-deficient conditions. Specifically, the D4625 apparatus (see Figure 1) was modified to allow a slow air purge of the bottle ullage during tests of a B100 and a diesel fuel. A comparison of these results is shown in Table 2. The results indicate that significantly higher levels of insolubles are obtained with the B100 when air purging is used. Because a real vehicle fuel tank will be agitated by vehicle motion and is open to the air, ullage purging will be employed in this 6-day test. This test is similar to a 24-hour storage stability test proposed by Bondioli and coworkers.[18]

Table 2. Results of D4625 (6-Day, 80°C Modification) Ullage Purge Scoping Tests

Sample identification	AL-27131-F[1]		AL-27070-F[2]	
Modified ASTM D4625[3]	with air flow	without air flow	with air flow	without air flow
filterable insolubles, mg/100 ml				
replicate 1	1.3	0.14	0.20	0.17
replicate 2	2.1	0.11	0.20	0.20
adherent insolubles, mg/100 ml				
replicate 1	1.7	0.80	0.31	0.14
replicate 2	2.0	0.80	0.20	0.20
total insolubles, mg/100 ml				
replicate 1	3.0	0.94	0.51	0.31
replicate 2	4.1	0.91	0.40	0.40
average total insolubles, mg/100 ml	3.6	0.93	0.46	0.36

[1] Soy-based B100
[2] Low-sulfur No. 2 diesel fuel
[3] 500-ml; 80°C; 1 week storage with and without air flow within ullage.

This test was performed on duplicate 500-ml samples. A 50-ml aliquot was removed from each bottle at the end of 1 week of storage. These were combined to make 100 ml and filtered to obtain a total insoluble result. The goals for this test are to:

- Determine whether there is an accelerated test that predicts simulated tank stability results.
- Assess the potential of a B100 accelerated stability measurement to predict blend stability in this scenario.
- Reveal how antioxidants can affect blend stability in the vehicle tank situation

High-Temperature Fuel System Environment

Biodiesel blends are subjected to temperatures as high as 150 C for short periods in the engine's fuel system. A substantial fraction of this fuel is recirculated to the vehicle fuel tank. Proposing a standard test that simulates this environment is somewhat more challenging. One possibility is ASTM D6468 (150 C and 180 minutes) but modified to make a gravimetric measurement of deposits. This test can be run on fresh fuel to simulate potential thermal stability in or around the injector (mentioned in the discussion of accelerated stability tests), or with aged fuel that has been in the fuel system for some time.

A worst-case scenario would be to test a fuel for 6 days at 80 C with the modified D4625, and then perform D6468. In this study one 50-ml aliquot was taken from each duplicate bottle after 1 week of storage. These were combined, and then split into two 50-ml samples for the D6468 test. Both 50-ml aliquots were filtered through the same filter to determine insolubles in mg/100 ml. The goals for this test are to:

- Determine whether there is an accelerated test that predicts simulated high-temperature zone stability.
- Assess the potential of a B100 accelerated stability measurement to predict blend high-temperature stability in this scenario.
- Reveal how antioxidants and other additives can effect high temperature stability.

This test, like the vehicle fuel tank simulation test, is not standard, and results should be interpreted with caution.

RESULTS: STABILITY OF B100

Accelerated Testing and Characterization

Table 3 lists the 19 B100 samples obtained, their feedstocks, and preliminary characterization results. These samples appear to cover the full range of feedstocks currently used in North America. Two samples failed the ASTM D6751 specification for biodiesel: one because of high acid value and the other because of high total glycerin. These samples are included in the accelerated stability tests, but were not considered for any additional testing in this study because of their poor quality. The samples that were selected for additional study were characterized much more extensively.

**Table 3. Biodiesel Samples Obtained and Preliminary Characterization
(values in bold exceed specification limits)**

Sample Identification	Feedstock	Total Acid Number, mg KOH/g	Total Glycerin, %(mass)	Free Glycerin, %(mass)
		ASTM D664	ASTM D6584	ASTM D6584
ASTM D6751-03a Limit:		0.80[(1)]	0.240	0.02
AL-27128-F	Canola	0.23	0.103	0.009
AL-27129-F	Palm Stearin	0.41	0.081	<0.001
AL-27137-F	Soy	0.05	0.144	0.002
AL-27138-F	Soy	0.33	0.016	0.002
AL-27140-F	Soy	0.20	0.022	0.015
AL-27141-F	Soy	0.13	0.121	0.005
AL-27142-F	Soy	0.07	0.216	0.003
AL-27144-F	Soy	0.39	0.221	0.004
AL-27145-F	Soy	0.49	0.192	0.005
AL-27146-F	Rapeseed	0.08	0.161	<0.001
AL-27148-F	Grease	0.69	0.121	<0.001
AL-27152-F	Rapeseed	0.09	0.15	0.001
AL-27153-F	Rapeseed	0.08	0.15	0.001
AL-27154-F	Grease	1.31	0.132	0.003
AL-27155-F	Soy	0.29	0.298	0.007
AL-27157-F	Soy	0.11	0.225	0.015
AL-27158-F	Soy	0.51	0.158	0.001
AL-27160-F	Tallow	0.46	0.188	0.002
AL-27161-F	Grease	0.37	0.151	0.003

[1] This limit was reduced to 0.50 mg KOH/g in mid-2006.

Oxidation stability data for the 19 B100 samples are tabulated in Appendix A. Figure 2 shows a histogram for OSI induction time measured for these samples. The samples show a broad distribution with results ranging from less than 1 to more than 9 hours. Figure 3 compares the OSI induction time results to ASTM D525 induction time results. Based on an earlier report, we anticipated that these tests would correlate to some extent.[17] The circled data points indicate that no oxidation was observed after 780 minutes, and the results are simply plotted as 780 minutes. Figure 3 indicates a poor correlation between these methods (r^2 of 0.4). Subsequent testing showed that D525 results were not predictive of B100 or blend performance. Consequently, D525 results are not discussed further.

Figure 4 shows a histogram for the ASTM D2274 deposit test. Ten of the 19 samples tested show deposits of less than 2 mg/100 ml; deposits for the other nine fuels cover the range to nearly 18 mg/100 ml. As observed previously,[19] the amount of adherent insolubles was 2 to 4 times higher than the amount of filterable insolubles. Figure 5 shows the inverse relationship between induction time and D2274 total insolubles. This is very similar to the relationship observed previously,[19] and in an approximate sense shows the expected inverse correlation. The filtrate from this test was also mixed with iso-octane to precipitate materials that are insoluble in nonpolar solvents. Fourteen of the samples exhibited less than 20 mg/100 ml; the balance ranged to 200 mg/100 ml. Figure 6 shows the relationship between total insolubles and iso-octane insolubles. The correlation is not perfect, but the only samples that produced very high levels of iso-octane insolubles also produced very high levels of total insolubles. This suggests that little new information is acquired by measuring iso-octane insolubles. Finally, the acid value of the filtered liquid was also measured and ranged from less than 1 to more than 5 mg KOH/g. Change in acid value over D2274 is closely correlated with total insolubles and inversely correlated with induction time, as shown in Figures 7 and 8.

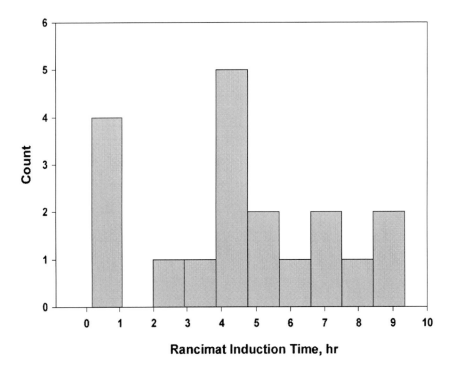

Figure 2. Histogram for B100 OSI or Rancimat induction time.

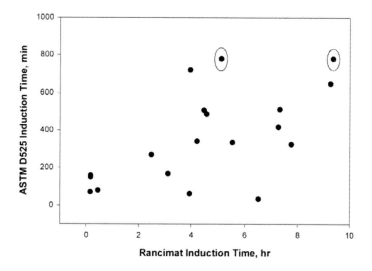

Figure 3. Relationship between B100 OSI or Rancimat induction time and ASTM D525 results. Circled data points indicate that no oxidation was observed after 780 minutes and the results are simply plotted as 780 minutes.

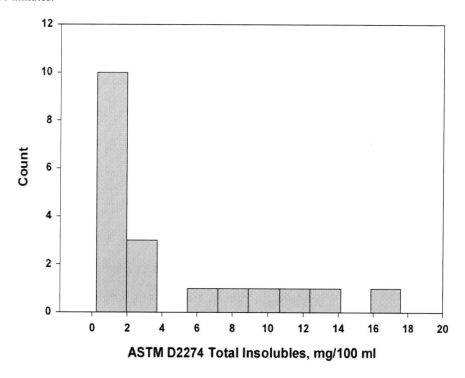

Figure 4. Histogram for B100 ASTM D2274 total insolubles.

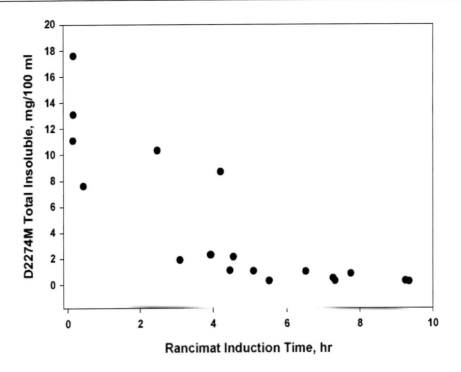

Figure 5. Relationship between B100 OSI or Rancimat induction time and D2274 total insolubles.

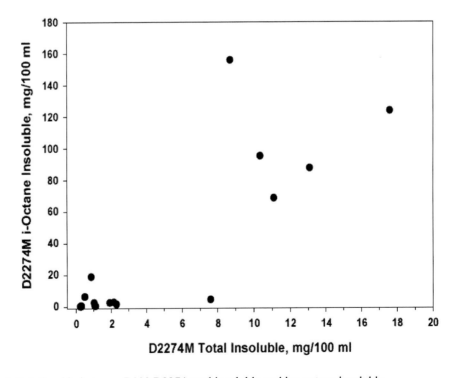

Figure 6. Relationship between B100 D2274 total insoluble and iso-octane insoluble.

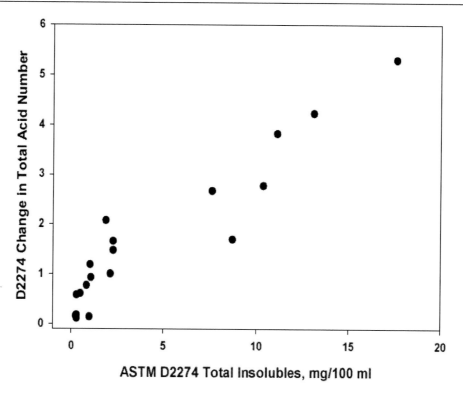

Figure 7. Change in total acid number on D2274 versus D2274 total insolubles for B100 samples.

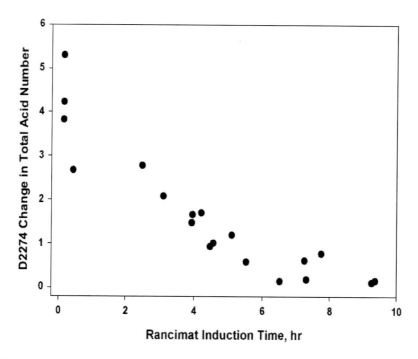

Figure 8. Change in total acid number on D2274 versus OSI or Rancimat induction time for B100 samples.

Table 4 lists results for the ASTM D6468 thermal stability test modified for gravimetric measurement of deposits. All samples produced low levels of deposits, and there is little discrimination among the samples.

Table 4. Results of ASTM D6468 Performed on B100 Samples at 150°C/180 Minutes and Modified for Gravimetric Determination of Deposits

Sample Identification	ASTM D6468 Modified
	Thermal Stability, deposit, mg
AL-27128-F	0.3
AL-27129-F	0.1
AL-27137-F	0.3
AL-27138-F	0.3
AL-27140-F	0.4
AL-27141-F	0.2
AL-27142-F	0.3
AL-27144-F	0.4
AL-27145-F	0.4
AL-27146-F	0.5
AL-27148-F	0.4
AL-27152-F	0.1
AL-27153-F	0.5
AL-27154-F	0.1
AL-27155-F	0.3
AL-27157-F	0.3
AL-27158-F	0.6
AL-27160-F	0.5
AL-27161-F	0.2

Based on these results, eight samples were selected for more detailed characterization, longterm storage tests (D4625 for 12 weeks), and for preparation of B5 and B20 blends. The downselection was based on covering the full range (high, medium, low) of values for each of the accelerated tests and on trying to include the full range of feedstocks. However, results for D6468 showed low levels of deposits (gravimetric) for all samples, so that test is not considered here. Additionally, the two B100 samples that failed to meet specifications for either acid value or total glycerin were not considered. The samples selected are shown in Table 5. Some compromises were made to meet all the study objectives, including covering all feedstocks. These eight samples were subjected to more extensive characterization tests. Results of these tests are shown in Table 6. All samples met the D6751 requirements for which they were tested.

Table 5. B100 Samples Selected for Long-Term Storage and Blending Studies

	Feedstock	OSI/Rancimat	D2274 Total	D2274 i-Octane
Observed Range		0.2–9.4 h	0.3–17.6 mg/100 ml	0.6–198 mg/100 ml
AL-27128-F	Canola	4.2 (med)	6.5 (med)	198 (high)
AL-27129-F	Palm stearin	3.1 (med)	1.9 (low)	2.6 (low)
AL-27137-F	Soy	6.5 (high)	1.0 (low)	2.6 (low)
AL-27138-F	Soy	0.5 (low)	7.6 (med)	4.4 (low)
AL-27141-F	Soy	5.5 (high)	0.3 (low)	0.6 (low)
AL-27148-F	Grease	7.8 (high)	0.9 (low)	19 (med)
AL-27152-F	Rapeseed	7.3 (high)	0.5 (low)	6.4 (low)
AL-27160-F	Tallow	0.2 (low)	17.6 (high)	124 (high)

Storage Stability of B100

Storage stability was assessed for the eight B100 samples by using ASTM D4625 (storage for 12 weeks at 43°C), numerical results are in Appendix B. For petroleum-based diesel fuels, each week of testing correlates to approximately 1 month of storage at ambient conditions. Figure 9 shows how induction times change over this test. All these biodiesel samples exhibited induction times shorter than 3 hours after 4 weeks, and all declined to very low levels at 8 weeks. Figure 10 shows peroxide value over the test and indicates that all but one of the samples had clearly degraded, even at 4 weeks. Figure 11 shows how acid value changes over the D4625 test. All but the most stable samples show a small increase in acidity at 4 weeks, but all still meet the 0.80 mg KOH/g limit that was in place when these samples were acquired. By 8 weeks (simulated 8 months) acidity had increased to above the 0.80 mg KOH/g limit for all samples. Figure 12 shows total insolubles formation over 12 weeks (simulated 8 months). These results show acid value and total insolubles in a range similar to that observed in previous investigations that used this test.[5,8,16] The level of B100 total insoluble on this test that is indicative of an unacceptable fuel blending component, and the repeatability of this test for B100, are unknown. The acid number increase may be a more accurate indicator than sediment formation of deleterious changes in B100 under these conditions. The data can mainly be interpreted by looking for a large increase in insolubles or by comparing the samples in a relative sense. Generally all the samples are beginning to experience significant increases in insolubles formation at the 12-week point. At 4 and 8 weeks the samples with induction times shorter than 3 hours have the highest levels of insolubles. Data for iso-octane insoluble and polymer content have a similar pattern (not shown), although polymer content remains low through 8 weeks and increases only at 12 weeks. Viscosity does not appear to be a sensitive indicator of oxidation for these samples.

Table 6. Characterization Results for B100 Samples Downselected for Further Study and Blending[*]

	Test Method	Units	AL-27128	AL-27129	AL-27137	AL-27138	AL-27141	AL-27148	AL-27152	AL-27160
Particulate Contamination	D6217, mod	mg/l	14.7	0.2	103.9	0.8	3.5	19.5	5.1	17.6
Total Water	D6304	ppm	656	217	149	562	131	118	298	1092
Flash Point	D93	°C	160	177	179	155	152	131	169	178
Elemental Analysis	D5185	ppm								
P			<1	<1	<1	<1	<1	<1	<1	<1
Na			<5	<5	<5	<5	<5	<5	5	5
K			<5	<5	<5	<5	<5	<5	<5	<5
Ca			1	<1	<1	<1	<1	<1	<1	2
Mg			<1	<1	<1	<1	<1	<1	<1	<1
Cu			<1	<1	<1	<1	<1	<1	<1	<1
Zn			<1	<1	<1	<1	<1	<1	<1	<1
Mg			<1	<1	<1	<1	<1	<1	<1	<1
Iso-Octane Insoluble	D4625, mod	mg/100 ml	3.9	0.1	1.9	1.1	2.3	0.1	0.1	0.5
Peroxide Value	D3703		217	105	12	50	98	9	44	17
Viscosity @ 40°C	D445		4.45	5.12	4.10	4.32	4.09	4.67	4.47	4.86
Polymer Content	ISO16931		2.63	0.17	0.82	0.22	1.03	1.46	2.17	4.91
Total Acid No.	D664	mg KOH/g	0.23	0.41	0.05	0.33	0.13	0.69	0.09	0.46
Free Glycerin	D6584	wt%	0.009	<0.001	0.002	0.002	0.005	<0.001	0.001	0.002
Total Glycerin	D6584	wt%	0.103	0.081	0.144	0.016	0.121	0.121	0.150	0.188

At the time these samples were collected the acid value was limited to 0.80 mg KOH/g maximum. In mid-2006 this limit was reduced to 0.50 mg KOH/g.

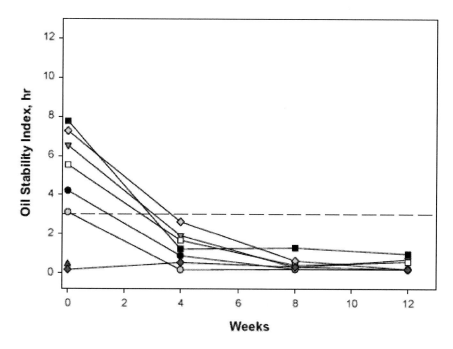

Figure 9. Change in OSI or Rancimat for B100 samples over 4 weeks in D4625 test.

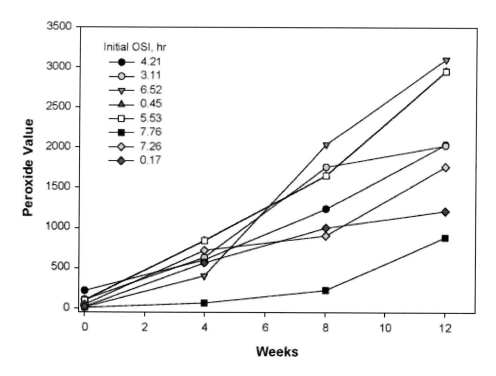

Figure 10. Peroxide value measured for B100 samples over the D4625 test.

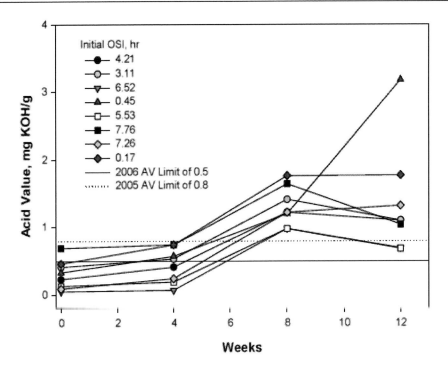

Figure 11. Change in acid value for B100 samples on the D4625 test.

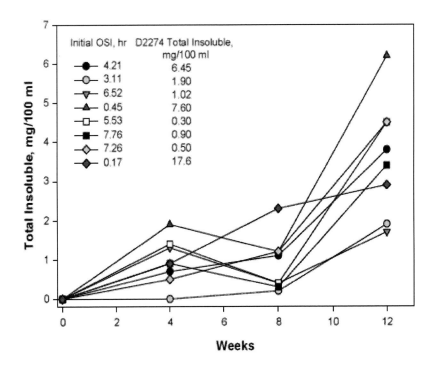

Figure 12. Total insoluble measured for B100 samples over the D4625 test.

B100 Antioxidant Testing

Biodiesel samples AL-27129-F and AL-27138-F were treated with two commercial antioxidants. Antioxidant treat rates as well as Rancimat induction times and total insolubles from ASTM D2274 are given in Table 7. Both additives are effective at increasing induction time and reducing insolubles formation for these samples. The biodiesels were selected to represent unstable material (AL-27138-F) with an induction time of 0.45 hours and moderately stable material (AL-27129-F) with an induction time of 3.11 hours.

Table 7. Results of Accelerated Stability Tests
or Biodiesel Treated with Antioxidant Additives

Sample Number	Description	Rancimat Induction	ASTM D2274		
			Total Insoluble	i-Octane Insoluble	Acid Value
		h	mg/100 ml	mg/100 ml	mg KOH/g
	AL-27138 with no additive	0.45	7.6	4.4	3.01
06-0077	AL-27138 plus 1000 ppm additive A	3.18	6.5	1.0	2.35
06-0078	AL-27138 plus 2000 ppm additive A	5.73	0.3	1.6	0.40
06-0079	AL-27138 plus 1000 ppm additive B	1.67	3.0	0.6	2.49
06-0080	AL-27138 plus 2000 ppm additive B	2.64	0.8	0.3	1.28
	AL-27129 with no additive	3.11	1.9	2.6	2.50
06-0081	AL-27129 plus 1000 ppm additive A	22.8	0.1	0.5	0.45
06-0082	AL-27129 plus 1000 ppm additive B	6.87	0.2	0.1	0.47

These samples were also tested for long-term storage stability with ASTM D4625; detailed results for these tests are reported in Appendix B. Results for total acid number are shown in Figure 13. The data indicate that both additives effectively prevented acid formation. Results for total insolubles are shown in Figure 14. Both antioxidants effectively suppressed insoluble formation in both biodiesel samples at both treat rates. Protection from insoluble and acid formation might have been achieved at lower treat rate.

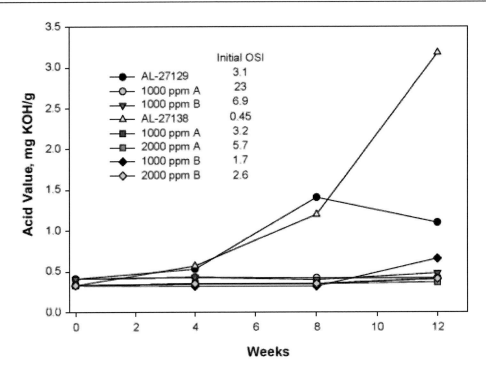

Figure 13. Total acid number results for D4625 long-term storage testing of antioxidant-treated B100 samples.

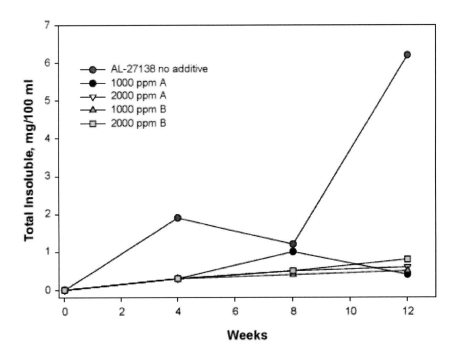

Figure 14. Total insoluble results of ASTM D4625 long-term storage test for antioxidant-treated B100 samples.

RESULTS: STABILITY OF BIODIESEL BLENDS

Diesel Fuel Properties and Stability

Characterization results for the six diesel fuel samples to be used for blending are shown in Table 8. Five are ULSD, and one is an on-road diesel fuel produced before ULSD was required. Based on the distillation T90, samples AL-27150F, AL-27166F, and AL-27176F are No. 1 diesel fuels; the others are No. 2 diesel fuels. Aromatic content is regarded as an important parameter for this study because fuels with higher aromatic content may be able to more readily solvate oxidized biodiesel molecules, which would otherwise precipitate as deposits in fuels with lower aromatic content. Total aromatics for the ULSD samples range from 8.2 to 22.1 mass percent. Because these ULSD were obtained in late 2005 they might not be representative of commercial ULSD in use today. All samples exhibited good stability on both D2274 and D6468.

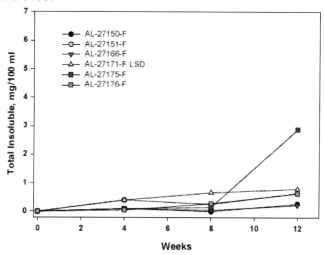

Figure 15. Total insolubles for petroleum diesel fuels on the D4625 test.

These fuels were also tested for long-term storage stability by using D4625 with measurement of insolubles formation. Results are shown in Figure 15 where all diesel samples were stable at 8 weeks and one sample, AL-27175, showed the initiation of degradation at 12 weeks. This is a No. 2 ULSD with the lowest aromatic content of the samples in this set. Detailed results for D4625 are listed in Appendix C.

The diesel fuels were also tested by using the fuel tank aging simulation of 1 week in the D4625 glass vessel at 80°C while the ullage was purged with air. Results are reported at the bottom of Table 8. These samples, after the 1 week of aging, were also tested with the D6468 thermal stability test but with gravimetric measurement of total insolubles. These results are also shown in Table 8. Because this test sequence is not standard, there is little information with which to compare these numbers. However, they do provide an important baseline for comparison with results for biodiesel blend samples.

Table 8. Characterization Results for Petroleum Diesel Samples To Be Used in Preparation of B5 and B20 Blends

Sample		ASTM D975 Limit (No. 2 Diesel)	AL27150F	AL27151F	AL27166F	AL27171F	AL27175F	AL27176F
ASTM D93	Flash point, °C	52	56	69	59	73	59	69
ASTM D5453	Sulfur, ppm	15 or 500	7.4	6.7	5.8	339.6	2.9	7.4
ASTM D86	T90, °C	282 min 338 max	274	313	269	319	333	236
ASTM D524	Carbon residue (on 10% distillation residue), mass%	0.35	0.07	0.04	0.06	0.13	0.05	0.08
ASTM D664	Acid number, mg KOH/g	none	0.01	0.03	0.01	0.01	0.01	0.01
ASTM D3703	Peroxide number	none	<1	<1	<1	<1	<1	<1
ASTM D2709	Water and sediment, vol%	0.05	0.01	0.01	0.01	0.01	0.01	0.01
ASTM D482	Ash content, mass%	0.01	<0.001	<0.001	<0.001	<0.001	<0.001	<0.001
ASTM D5186	Total Aromatics, mass%	none	15.7	22.1	18.1	36.2	8.2	19.3
	Monoaromatics, mass%	none	14.4	19.9	17.1	27.6	7	17.4
	Polynuclear aromatics, mass%	none	1.3	2.1	1	8.7	1.2	1.9
ASTM D6217	Particulate contamination, mg/l	none	0.5	0.4	0.8	0.8	1.2	0.3
ASTM D2274	Total Insolubles, mg/100 ml	none	0.25	0.25	0.5	0.2	0.1	0.05
ASTM D6468	Thermal stability, 150°C/180 min % reflectance	none	100	100	100	98	95	100
ASTM D4625	Modified, 80°C, air purge, 1 week	none						
	Adherent insoluble, mg/100 ml		0.3	0.4	0.1	0.1	0.2	0.2
	Filterable insoluble, mg/100 ml		0.7	0.2	0.1	0.1	0.4	0.2
	Total insoluble		1.1	0.6	0.2	0.2	0.6	0.4
ASTM D6468	Modified, 150°C/180 min, gravimetric on fuel from D4625, 1 week 80°C, mg/100 ml	none	2.2	0.2	0.1	0.1	0.4	0.8

Stability of B5 Blends

Accelerated Tests

Six diesel fuels were acquired, and eight B100 samples were selected from the original 19. These were used to prepare 48 B5 blends that have been tested by using the following accelerated stability tests:

- OSI/Rancimat induction time
- ASTM D2274 (biodiesel modification)
- ASTM D6468 percent reflectance and gravimetric modification.

Results for accelerated tests with B5 blends are in Appendix D. Figure 16 shows a histogram of the Rancimat induction time results. The samples fall into three categories: short, medium, and long induction times. Figure 17 shows the histogram for total insolubles measured on the D2274 test. Most of the samples are reasonably stable, with less than 2.5 mg/100 ml. There are also samples with intermediate and high levels of insolubles. Figures 18 and 19 show the reflectance and gravimetric results from the D6468 thermal stability test. Most samples are thermally stable, although several produce significant amounts of insolubles.

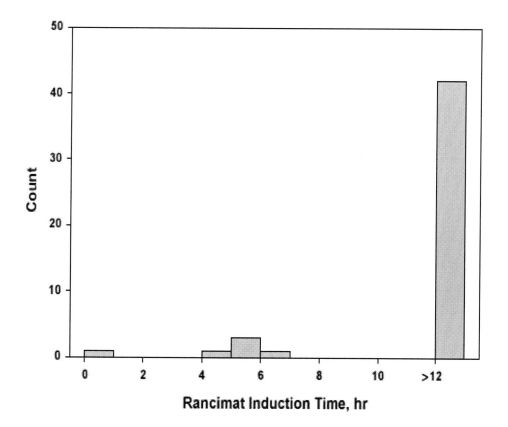

Figure 16. Histogram of OSI or Rancimat induction times for B5 samples.

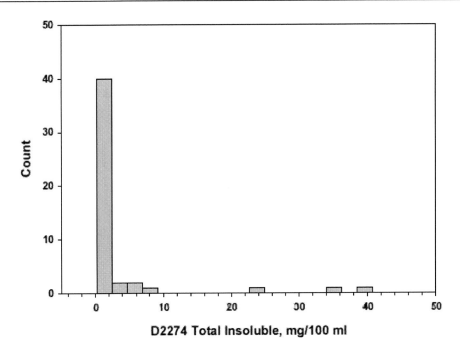

Figure 17. Histogram of D2274 (biodiesel modification) total insolubles results for B5 samples.

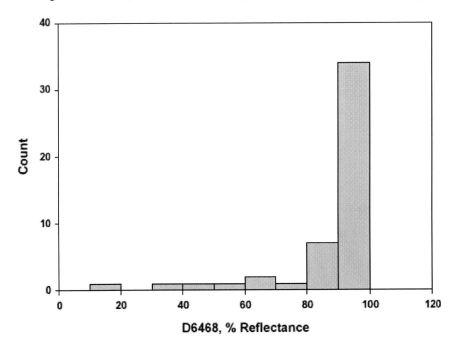

Figure 18. Histogram of D6468 percent reflectance results for B5 samples.

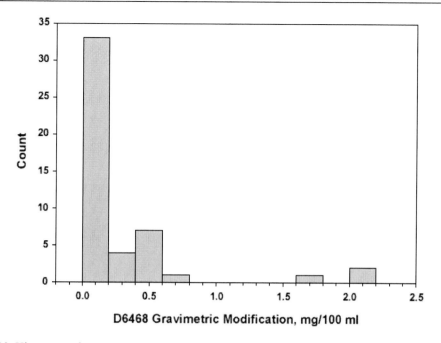

Figure 19. Histogram of D6468 gravimetric insoluble results for B5 samples.

Based on these distributions, and so as not to exclude any one biodiesel or diesel fuel from the group, the B5 blend samples listed in Table 9 were selected for more detailed study. This included long-term storage testing, as well as tests to simulate aging in the vehicle fuel tank and degradation in the high-temperature environment of a diesel engine fuel injection system. Biodiesel AL-27138, an unstable material, produced a B5 blend with a short induction time and high D2274 insolubles, but is thermally stable. Biodiesel AL-27160, also unstable, produced a B5 blend with a short induction time and low D2274 deposits, but appears to be thermally unstable. These data confirm previous results and indicate that biodiesel and B5 blends are generally thermally stable,[16] and with the implementation of the 3-hour induction period there appears to be no additional information to be gained about blends from the thermal stability values of the B100.

Table 9. B5 Samples Downselected for More Detailed Study

Sample	B100	Diesel	Rancimat h	D2274 mg/100 ml	D6468 % Reflectance/Insoluble mg/100 ml
CL06-200	27128	27166	>12	5.5	98/0.1
CL06-215	27129	27176	>12	1.5	99/0.1
CL06-227	27137	27175	>12	0.3	94/0.1
CL06-235	27138	27150	4.8	40.7	88/0.1
CL06-252	27141	27175	>12	0.3	96/0.1
CL06-264	27148	27175	>12	0.4	93/0.2
CL06-273	27152	27151	>12	0.3	99/0.1
CL06-287	27160	27171	5.5	0.8	19/2.0

Several B5 samples were also treated with antioxidant additives and evaluated on accelerated tests. Results are shown in Table 10. Most of the B5 samples are fairly stable without additive on these tests. Antioxidants caused a large reduction in D2274 total insolubles for the one sample that exhibits significant instability and increased the induction time for all samples. Additive treat rates given are for addition to the B100 before it is blended.

Table 10. Accelerated Test Results for B5 Samples Treated with Antioxidant

Sample Number	Description	Rancimat Induction Time h	ASTM D2274 Total Insoluble mg/100 ml
06-236	5% AL-27138 in AL-27151 no additive	5.8	23.5
06-548	5% AL-27138 plus 1000 ppm additive A in AL-27151	9.0	–
06-331	5% AL-27138 plus 2000 ppm additive A in AL-27151	>12	0.3
06-332	5% AL-27138 plus 1000 ppm additive B in AL-27151	>12	0
06-333	5% AL-27138 plus 2000 ppm additive B in AL-27151	>12	0
06-211	5% AL-27129 in AL-27151 no additive	>12	1.0
06-334	5% AL-27129 plus 1000 ppm of additive A in AL-27151	>12	0
06-550	5% AL-27129 plus 1000 ppm of additive B in AL-27151	>12	–
06-214	5% AL-27129 in AL-27175 no additive	>12	0.6
06-552	5% AL-27129 plus 1000 ppm of additive B in AL-27175	>12	–
06-239	5% AL-27138 in AL-27175 no additive	6.9	0.2
06-552	5% AL-27138 plus 1000 ppm of additive A in AL-27175	>12	0.2
06-516	5% AL-27138 plus 2000 ppm additive A in AL-27175	>12	0.2
06-518	5% AL-27138 plus 1000 ppm additive B in AL-27175	>12	0
06-520	5% AL-27138 plus 2000 ppm additive B in AL-27175	>12	0.3

Storage Stability

The eight B5 blends listed in Table 9 were tested for long-term storage stability with ASTM D4625. Numerical results are given in Appendix E. Figure 20 shows results for peroxide value (PV). At 4 weeks, the peroxide value results indicate that some degradation is occurring for samples produced from B100 with an induction time shorter than 1 hour, and for one sample with a longer induction time. However, at 8 weeks all samples show low peroxides, and at 12 weeks the only sample with high peroxides was produced from an unstable B100. Acid value increase for B5 blends over the D4625 test is shown in Figure 21. The only B5 sample in which acid value increased significantly was blended from a B100 with an induction time of 0.5 hour. Total insoluble formation is shown in Figure 22. Again, the only sample in which insolubles increased significantly was blended from a B100 with an induction time of 0.5 hour (AL-27138). Figure 23 shows results for several B5 blends that were treated with antioxidants. Treatment of B5 blends containing AL-27138 eliminated the increase in insolubles observed at 12 weeks without antioxidant, with all results below 2 mg/100 ml. Antioxidant treatment also stabilized the induction time for all samples at more than 10 hours over the 12-week test.

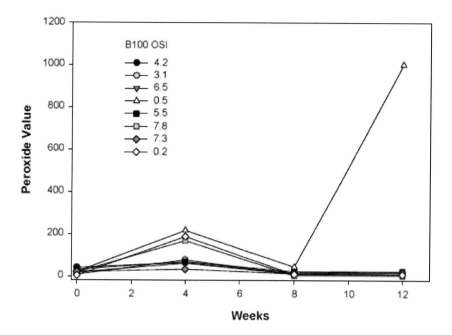

Figure 20. Peroxide value over the D4625 test for B5 samples.

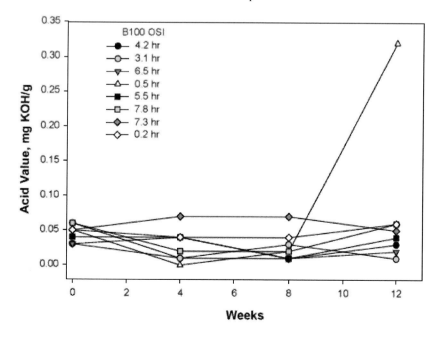

Figure 21. Acid value for B5 blends over the D4625 test.

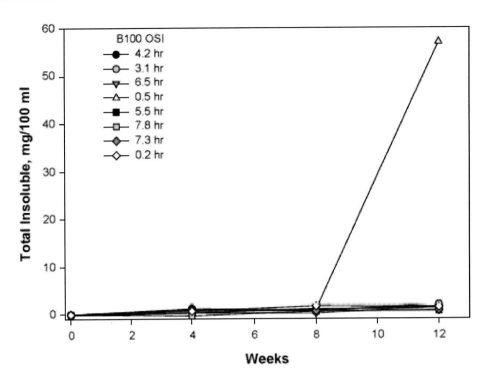

Figure 22. Total insolubles formation for B5 blends in the D4625 test.

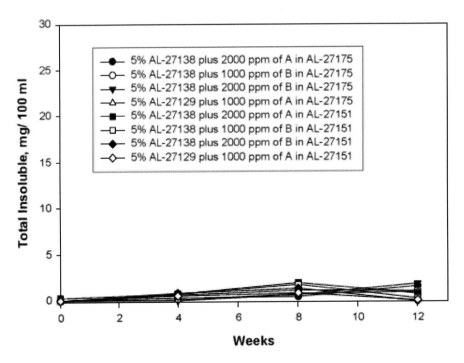

Figure 23. Total insolubles formation for B5 samples treated with antioxidant additives on the D4625 test.

Simulated Vehicle Fuel Tank Aging As described in the methods section, fuel tank aging was simulated by storing the biodiesel blend in the D4625 flask at 80°C for 1 week with ullage purge, followed by measurement of peroxide number, acid number, and insolubles. Figures 24, 25, and 26 show results for peroxide, acid, and total insolubles, respectively, as a function of the B100 induction time. Detailed results are given in Appendix E. For all three measures, significant degradation occurs only for B100 induction times shorter than 1 hour. Similar plots could be made as a function of the B5 induction times (not shown) and indicate that degradation is not observed for B5 induction times longer than 6 hours. Although also not shown, antioxidant additives reduced insolubles formation on this test for the one biodiesel sample that generated high levels of insolubles.

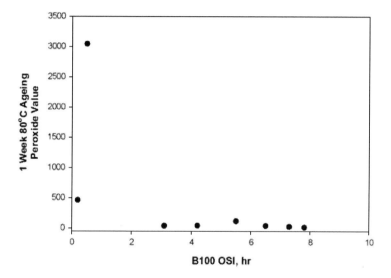

Figure 24. Peroxide value for B5 blends after 1 week aging at 80°C, as a function of B100 OSI or Rancimat induction time.

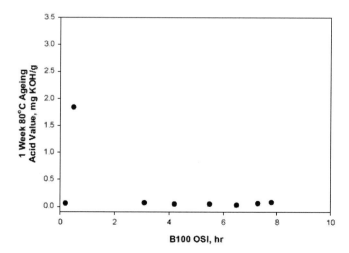

Figure 25. Acid value for B5 blends after 1 week aging at 80°C, as a function of B100 OSI or Rancimat induction time.

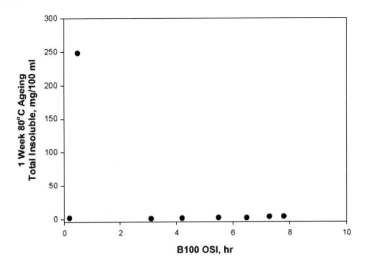

Figure 26. Total insoluble for B5 blends after 1 week aging at 80°C, as a function of B100 OSI or Rancimat induction time.

Simulated Vehicle Tank Aging Followed by High Temperature Testing To simulate the stability in the high-temperature environment of a diesel engine fuel injection system, we subjected the samples that had been aged for 1 week at 80°C (see previous section), then conducted the D6468 thermal stability test (150°C for 180 minutes). In addition to measuring filter reflectance, the amount of deposits on the filter was also quantified gravimetrically by using procedures that are identical to those in D2274. Results are shown in Figure 27 as a function of the blending B100 induction time. Detailed results are listed in Appendix E. Low values of reflectance or high values of gravimetric insoluble occur only for B100 with induction times shorter than 1 hour. Treatment of this B100 with antioxidants eliminated the formation of insolubles for this sample (not shown). Notably, the B5 blend produced from AL-27160, which demonstrated poor thermal stability on the D6468 test before aging, does not generate high deposits on this test after 1 week of aging.

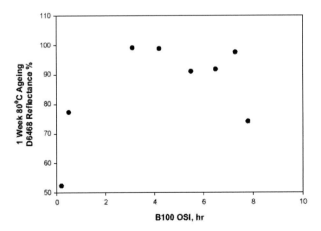

Figure 27. Filter reflectance and gravimetric insolubles formation for B5 samples aged for 1 week at 80°C and tested with the D6468 thermal stability test as a function of B100 OSI or Rancimat induction time.

Stability of B20 Blends Accelerated Tests Six diesel fuels were acquired, and eight B100 samples were selected from the original 19. These were used to prepare 48 B20 blends that were tested with the following accelerated stability tests:

- OSI/Rancimat induction time
- STM D2274 (biodiesel modification)
- ASTM D6468 percent reflectance and gravimetric modification.

Detailed results are listed in Appendix F. Figure 28 shows a histogram of the Rancimat induction time results. Figure 29 shows the histogram for total insolubles measured on the D2274 test. Most of the samples are stable, with less than 2.5 mg/100 ml. There are also samples with intermediate and high levels of insolubles. Figures 30 and 31 show the reflectance and gravimetric results from the D6468 thermal stability test. Only one sample exhibited poor D6468 reflectance values (below 70, the generally accepted level for petrodiesel). Although no generally accepted gravimetric value has been established for D6468, most values were below 1 mg/100 ml, and four samples gave higher results. There were fewer poor samples for D6468, but the sample set did contain samples that could be considered both stable and unstable on all the stability tests.

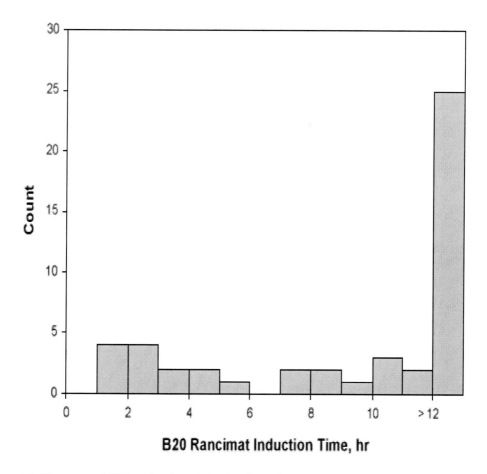

Figure 28. Histogram of OSI or Rancimat induction times for B20 samples.

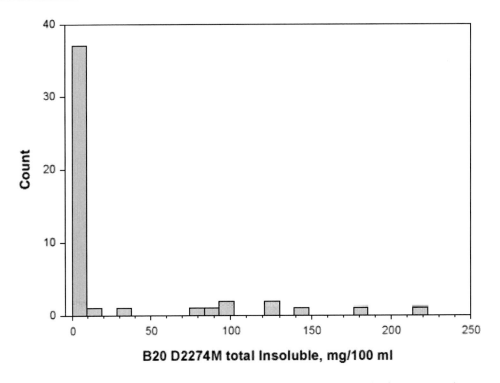

Figure 29. Histogram of D2274 (biodiesel modification) total insolubles results for B20 samples.

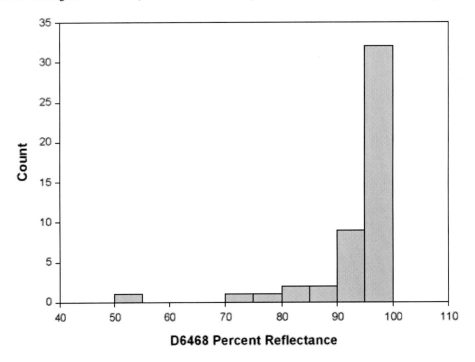

Figure 30. Histogram of D6468 percent reflectance results for B20 samples.

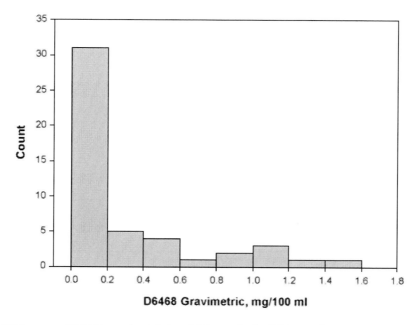

Figure 31. Histogram of D6468 gravimetric insoluble results for B20 samples.

Based on these distributions, and so as not to exclude any one biodiesel or diesel fuel from the group, we selected the samples listed in Table 11 for further testing of long-term storage stability, stability conditions simulating a vehicle fuel tank, and conditions simulating the high-temperature conditions of a diesel engine fuel injection system.

Table 11. B20 Samples Downselected for More Detailed Study

Sample	B100	Diesel	Rancim at h	D2274 mg/100 ml	D6468 % Reflectance/Insolublesmg/ 100 ml
CL06-196	27128	27175	7.8	0.3	95/0.1
CL06-204	27129	27150	>12	0.4	98/0.1
CL06-237	27137	27171	>12	0.6	82/0.9
CL06-242	27138	27151	1.8	124.8	99/0.1
CL06-256	27141	27166	10.6	0.4	99/0.1
CL06-268	27148	27166	>12	0.3	97/0.2
CL06-282	27152	27175	>12	1.2	97/0.1
CL06-295	27160	27176	4.1	33.7	72/0.1

Several additional B20 samples prepared with antioxidant treated B100 were also tested for stability. Results are shown in Table 12. The antioxidants generally reduced insoluble levels effectively and increased induction time for the samples prepared from the highly unstable B100 AL-27138. Additive treat rates given are for adding antioxidant to the B100.

Table 12. Accelerated Test Results for B20 Samples Treated with Antioxidant

Sample Number	Description	Rancimat Induction Time h	ASTM D2274 Total Insolubles mg/100 ml
06-242	20% AL-27138 in AL-27151 no additive	1.8	124.8
06-549	20% AL-27138 plus 1000 ppm additive A in AL-27151	4.4	–
06-335	20% AL-27138 plus 2000 ppm additive A in AL-27151	9.4	0.5
06-336	20% AL-27138 plus 1000 ppm additive B in AL-27151	4.6	0.3
06-337	20% AL-27138 plus 2000 ppm additive B in AL-27151	7.0	0.1
06-205	20% AL-27129 in AL-27151 no additive	>12	0.3
06-338	20% AL-27129 plus 1000 ppm of additive A in AL-27151	>12	0.2
06-551	20% AL-27129 plus 1000 ppm of additive B in AL-27151	>12	–
06-208	20% AL-27129 in AL-27175 no additive	>12	0.3
06-553	20% AL-27129 plus 1000 ppm additive B in AL-27175	>12	–
06-245	20% AL-27138 in AL-27175 no additive	1.7	222.8
06-523	20% AL-27138 plus 1000 ppm additive A in AL-27175	6.0	1.7
06-517	20% AL-27138 plus 2000 ppm of additive A in AL-27175	>12	0.2
06-519	20% AL-27138 plus 1000 ppm additive B in AL-27175	4.5	0.5
06-521	20% AL-27138 plus 2000 ppm additive B in AL-27175	6.6	0.2

Storage Stability

The eight B20 blends listed in Table 11 were tested with ASTM D4625 for long-term storage stability. Numerical results are given in Appendix G. Figure 32 shows results for peroxide value. The two unstable B100 samples, with induction times shorter than 1 hour, produced B20 blends that clearly degraded on this test. Acid value increase for B20 blends over the D4625 test is shown in Figure 33. B20 prepared from unstable B100 shows a significant increase in acidity. One other B100 (with an induction time of 6.5 hours) shows a small acidity increase, but this sample shows no increase in peroxide, so it may not actually be degrading. The acid number increase for this sample is also close to the reproducibility of the test method. Total insolubles formation is shown in Figure 34. The only sample that increased significantly in insolubles was blended from a B100 sample with an induction time of 0.5 hours (AL-27138). One B100 sample with a short induction time (0.2 hours) did not exhibit insolubles formation in B20 over the 12-week D4625 test. This indicates that the B100 induction period may be a conservative measurement of blend performance as the data indicate not all B100 failing the 3-hour induction period would form deleterious amounts of sediment when aged under these conditions.

Figure 35 shows results for several B20 blends treated with antioxidants, primarily B20 produced from the unstable B100, AL-27138. None of the antioxidant-treated samples showed a significant increase in insolubles formation. Antioxidant treatment also stabilized the induction time for these samples such that there was little change over the 12-week test. This result could indicate another potential solution to addressing stability with biodiesel and biodiesel blends, that being the mandatory addition of antioxidants. This would require further study, but confirmed use of antioxidants might be a better control strategy than any of the tests proposed here.

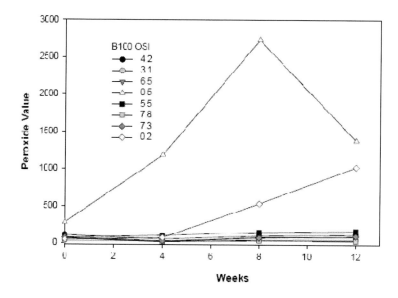

Figure 32. Peroxide value for B20 samples tested on D4625 test.

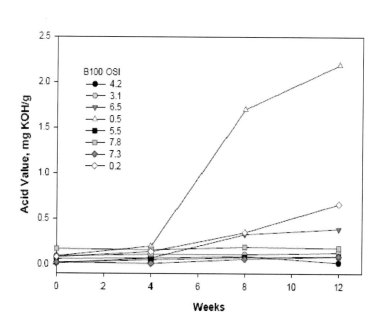

Figure 33. Acid value for B20 samples tested on the D4625 test.

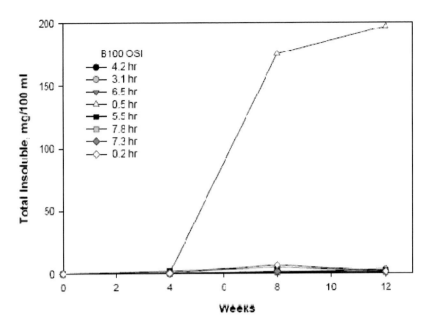

Figure 34. Total insolubles formed from B20 samples tested on the D4625 test.

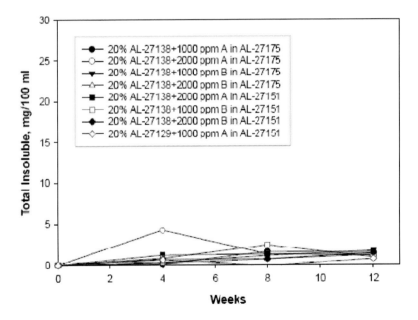

Figure 35. Total insolubles formation for B20 samples treated with antioxidant additives on the D4625 t
est.

Simulated Vehicle Fuel Tank Aging

As described in the methods section, worst-case fuel tank aging was simulated by storing the biodiesel blend in the D4625 flask at 80°C for 1 week with ullage purge, followed by measurement of peroxide number, acid number, and insolubles. This is not a standard test, and results should be interpreted with caution. Detailed results are given in Appendix G. An increase in peroxides is observed for all these B20 samples under these severe test conditions, as shown in Figure 36; however, the largest increase in acid value was observed for blends prepared from B100 with induction times shorter than 1 hour (Figure 37). B20 induction time is a somewhat better predictor of the final acidity on this test, as shown in Figure 38. The B100 induction time was not predictive of insolubles formation on this test for B20 samples as shown in Figure 39, nor was B20 induction time or B100 D2274 total insolubles (not shown). Antioxidant additives effectively reduced total insolubles formation on this test (not shown), although in some cases the additive treat rate may have been too low to completely eliminate insolubles (see Appendix G).

Because we cannot correlate results on this test to what actually happens in a vehicle fuel tank, the levels of insolubles shown in Figure 39, and the levels of acidity developed for the more stable B100 in Figure 37, may not indicate that these fuels would cause vehicle operational or durability problems. Interpretation would be more straightforward if we observed large differences for less stable versus more stable B100, as was the case for B5 blends. Neither anecdotal nor detailed field data on B20 show a propensity for significant vehicle fuel filter clogging, which should be the case if sediment actually forms at levels similar to this 1-week aging test in the field.[20]

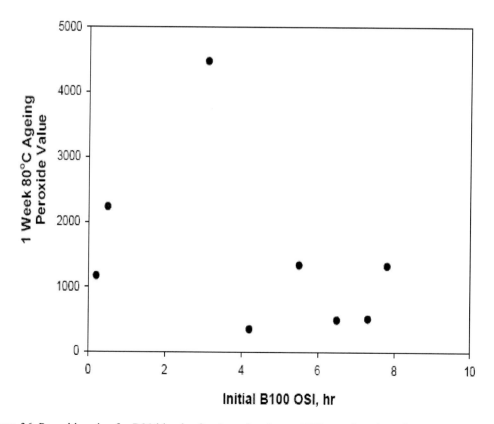

Figure 36. Peroxide value for B20 blends after 1 week aging at 80°C, as a function of B100 OSI or Rancimat induction time.

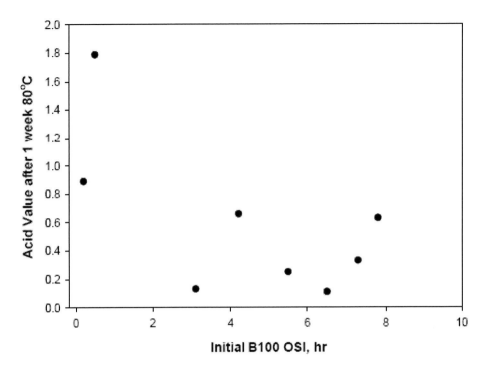

Figure 37. Acid value for B20 blends after 1 week aging at 80°C, as a function of B100 OSI or Rancimat induction time.

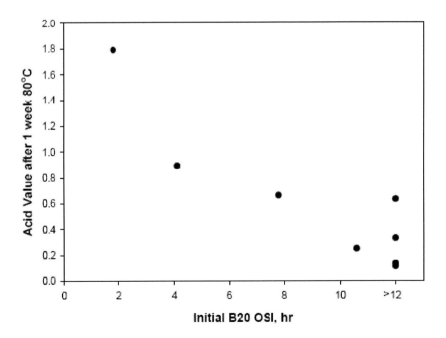

Figure 38. Acid value for B20 blends after 1 week aging at 80°C, as a function of B20 OSI induction time.

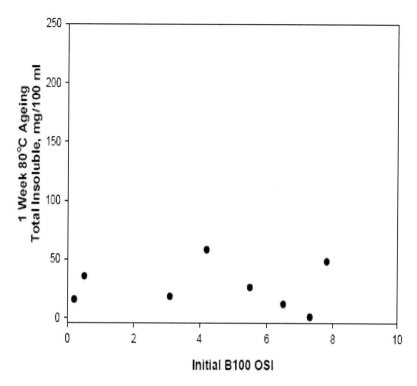

Figure 39. Total insolubles for B20 blends after 1 week aging at 80°C, as a function of B100 OSI induction time.

Simulated Vehicle Tank Aging Followed by High-Temperature Testing

Worst-case stability in the high-temperature environment of a diesel engine fuel injection system is simulated by subjecting the samples that had been aged for 1 week at 80°C (see previous section) to the D6468 thermal stability test (150°C for 180 minutes). We measured filter reflectance, and used procedures identical to those in D2274 to gravimetrically quantify the amount of deposits on the filter. Results are shown in Figure 40 as a function of the blending B100 induction time. Detailed results are listed in Appendix G. Low reflectance values are not observed for any of the B20 samples. High values of gravimetric insolubles occur only for B100 with induction times shorter than about 4 hours. Treating these B100 blends with antioxidants eliminated the formation of insolubles on this test (not shown). Notably, the B20 blend produced from AL-27160, which demonstrated marginal thermal stability on the D6468 test before aging, does not generate high deposits on this test after 1 week of aging. Similar results were observed for the B5 blend prepared from this B100.

As noted for the fuel tank simulation test, because we have no correlation of this test sequence (1 week at 80°C followed by D6468 thermal stability testing) with performance in actual engine fuel systems, whether any of the fuels tested would actually cause engine operational or durability problems is unclear. The B100 induction time specification of 3 hours does, however, eliminate the highest sediment-forming blend under these test conditions. But overall this testing approach does not provide an obvious delineation between fuels that are stable and those that are not under these conditions.

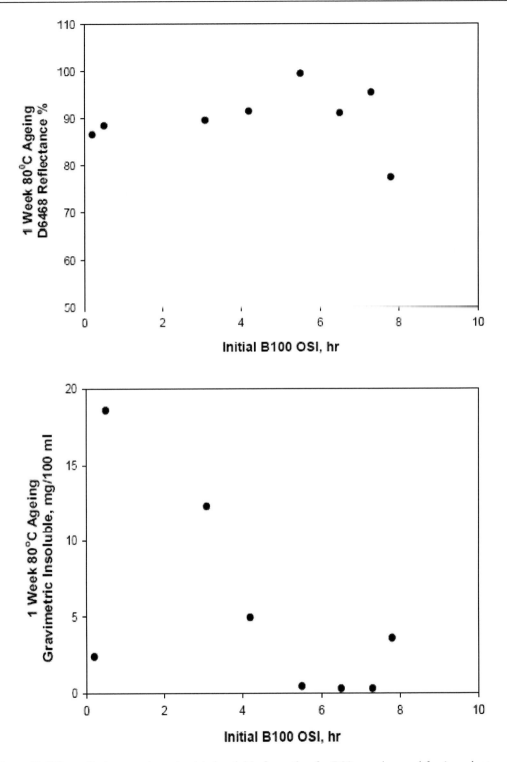

Figure 40. Filter reflectance and gravimetric insoluble formation for B20 samples aged for 1 week at 80°C and tested in the D6468 thermal stability test.

DISCUSSION

Do Accelerated Tests Predict B100 Storage Stability?

The D4625 storage data for B100 at 8 and 12 weeks indicate that none of the accelerated tests are good—or even average—predictors of the degree of biodiesel degradation on this test at these storage times. Almost all samples at both 4 and 8 weeks had sediment formation below 2 mg/100 ml, which indicates fairly low sediment formation on this test, but all samples exhibited an increase in acid number at 8 weeks to out-of-specification levels. Attempts to correlate biodiesel degradation at 4 weeks with results of initial stability tests were slightly more successful. Figure 41 shows that initial induction time is a rough predictor ($r_2 = 0.55$) of induction time at 4 weeks. Figure 42 suggests that if the initial degree of oxidative degradation (as measured by the peroxide number or value) is low, the peroxide number will be low at 4 weeks. However, the curve is very steep at low peroxide levels. Nevertheless, this confirms that exposure of the biodiesel to oxidative conditions that cause peroxide formation should be avoided. Figure 43 shows the change in acid value over 4 weeks as a function of the initial induction time. Samples with longer induction times in general experience a smaller change in acid value, although clearly other factors affect the results. Operating on the idea that the initial degree of degradation (peroxide number) combined with some measure of stability (induction time or D2274 insoluble) might be a better predictor, we performed simple multiple regression analysis, but r_2 values for these regressions were quite low.

Real-world experience and common sense suggest that samples with longer induction times or lower levels of D2274 deposits are more stable, but the results of these accelerated tests are not strong predictors of storage stability on the D4625 test at 4, 8, or 12 weeks. One possibility not examined here is that even at 4 weeks these samples have degraded so far that their initial state is no longer predictive of their condition. Storage results for a shorter period (2 weeks, for example) might yield stronger correlations. However, the data in Figure 9 show that if induction time is near or below the 3-hour limit, B100 will most likely fail to meet its specification for stability before 4 months. Even samples with induction times longer than 6 hours are out of specification for oxidation stability at 4 weeks on this test, suggesting that any B100 not treated with antioxidant should not be stored for more than 2 or 3 months. The B100 could be stored significantly longer before it goes out of specification for acid value or before significant levels of deposits form.

The D4625 test condition of 43°C may not be appropriate for B100. This test was developed to provide slightly accelerated conditions for assessing the stability of petroleum-derived fuels. For these fuels reactions that occur at typical storage temperatures are apparently the same as those that occur at 43°C. This may not be the case for B100. For example, Bondioli and coworkers[18] compared the results from D4625 tests[8] and storage for 1 year in drums at ambient conditions.[7] They observed only modest degradation in the 1-year storage study, but for D4625 storage they saw significant increases in peroxide value and insolubles at only 4 weeks, along with significant increases in acidity at 8 to 12 weeks. These results suggest that degradation reactions that occur in B100 are accelerated much more by increasing the temperature to 43°C than are those that occur in conventional diesel fuels at this temperature.

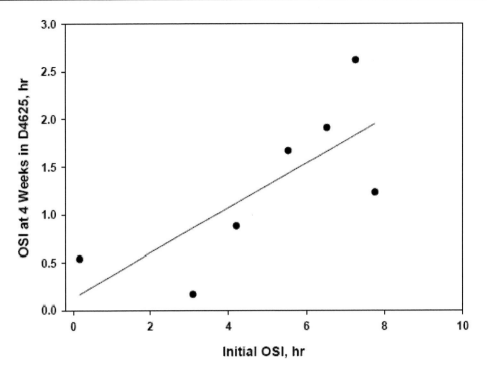

Figure 41. B100 OSI or Rancimat induction period at 4 weeks on the D4625 test as a function of initial induction time.

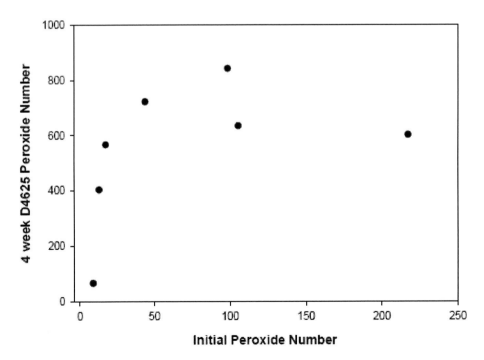

Figure 42. Peroxide number at 4 weeks on the D4625 test for B100 samples, as a function of initial peroxide number.

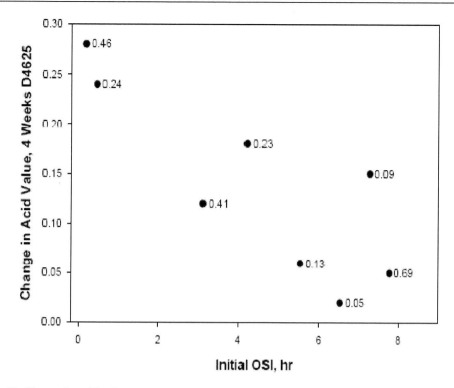

Figure 43. Change in acid value at 4 weeks on D4625 as a function of initial OSI or Rancimat induction time for B100 samples. Data labels indicate initial acid value.

Is B100 Stability a Predictor of Blend Stability?

Accelerated Tests

The accelerated test data indicate that B100 stability is, in some sense, a good predictor of blend stability for both B5 and B20 blends. For B5 blends the B5 induction time as a function of the B100 induction time is shown in Figure 44. The experiment was ended at 12 hours for B5 blends. If the B100 has an induction time of about 3 hours or longer, the B5 blend is guaranteed to have a 12-hour or longer induction time. A similar plot for B5 D2274 total insolubles is shown in Figure 45. Again, if the B100 has an induction time longer than about 3 hours, the B5 blend will always have D2274 deposits of less than 2.5 mg/100 ml. For less stable B100, high levels of insolubles—in fact, much higher than those observed for the B100 used to make the blend—can form. Another way to look at the data is to compare B5 D2274 total insolubles with B100 D2274 total insolubles as shown in Figure 46. If the B100 total insolubles are less than about 6 mg/100 ml, the B5 blend produces total insolubles below 2.5 mg/100 ml. The 2.5 mg/100-ml level was chosen because it is used as a stability requirement for diesel fuels in several pipeline systems.

Figure 47 compares B20 induction time to B100 induction time. For all but one of 48 samples, if the B100 induction time is longer than about 3 hours, the B20 induction time will exceed 6 hours. Figure 48 shows that if B100 induction time is longer than about 3 hours, the B20 D2274 total insolubles will be below 2.5 mg/100 ml. If B100 induction time is shorter than 3 hours, high levels of insolubles can form, much higher than those observed for the

B100 that was used to make the blend. Figure 49 shows that if B100 D2274 total insolubles are below about 6 mg/100 ml, B20 total insolubles will be below 2.5 mg/100 ml.

Thus, accelerated tests do not show a direct proportionality or correlation between B100 stability and blend stability, but if the B100 stability is above a certain level (roughly a 3-hour induction time), blends prepared from that B100 appear to be stable. Based in part on these results, ASTM has recently included a 3-hour minimum induction time requirement in the D6751 specification for B100 (this was originally included in D6751-06b).

Fang and McCormick recently reported on a negative synergism or antagonism for deposit formation from biodiesel blends on the D2274 test.10 As noted in the Background chapter, blends in the B20 to B30 range produced much higher levels of deposits than would be predicted from a weighted average of the results obtained for the diesel and biodiesel blending components. Figures 45 and 48 also show this effect, here even for B5 blends, although deposit levels are much higher for B20 blends.

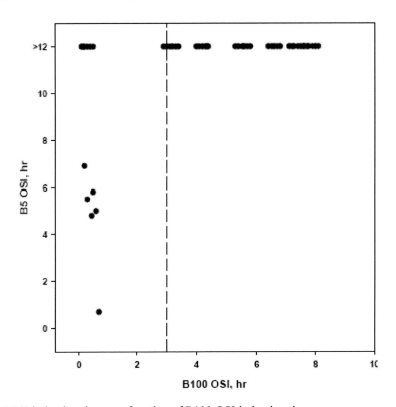

Figure 44. B5 OSI induction time as a function of B100 OSI induction time.

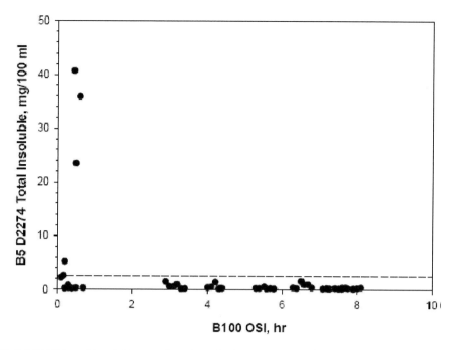

Figure 45. B5 D2274 total insoluble as a function of B100 OSI or Rancimat induction time.

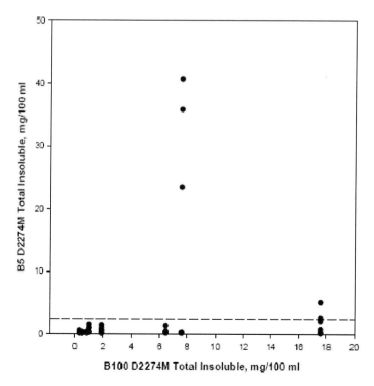

Figure 46. B5 D2274 total insoluble as a function of B100 D2274 total insoluble.

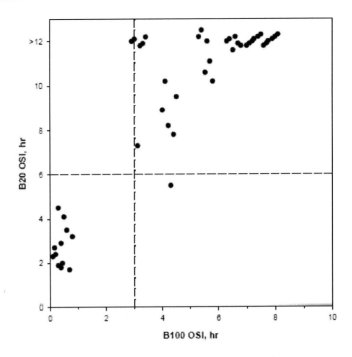

Figure 47. B20 OSI induction time as a function of B100 OSI induction time.

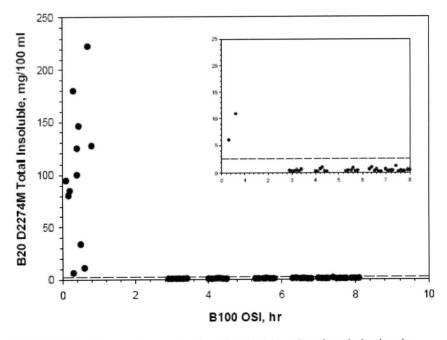

Figure 48. B20 D2274 total insoluble as a function of B100 OSI or Rancimat induction time.

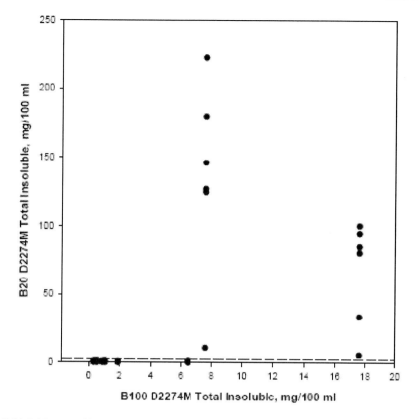

Figure 49. B20 D2274 total insoluble as a function of B100 D2274 total insolubles.

Aging and In-Use Scenario Tests
(Storage, Vehicle Tanks, High-Temperature Environment)

These tests include the D4625 12-week storage test, the 1 week at 80°C test, and the subsequent high-temperature thermal stability test. B100 stability on accelerated tests seems to be a reasonable predictor of stability of B5 and B20 blends on the D4625 test. As shown in Figures 20, 21, and 22; the only B5 blend that exhibited oxidative deterioration on this test was prepared from a B100 with an induction time of 0.5 hours. Similarly for B20, Figures 32, 33, and 34 show that only B20 prepared from B100 with short induction times (in this case 0.5 and 0.2 hours) exhibited oxidative degradation. So, although questions remain about the suitability of the 43°C storage test for B100 and about whether 1 week of storage on this test for biodiesel and biodiesel blends equates to 1 month of real-world storage (as is generally assumed for petroleum-derived fuels), biodiesel blends prepared from B100 with 3-hour or longer induction times do not appear to degrade under these conditions.

For the worst-case simulated fuel tank aging test (1 week at 80°C) Figures 24, 25, and 26 show significant oxidative degradation only for B5 blends prepared from a B100 with 0.5 hour induction time. The picture is less clear cut for B20, where Figures 36 through 39 indicate little relationship between B100 induction time and oxidative degradation under these conditions. However, B20 induction time was reasonably predictive of the acid value (and acid value increase) on this test. Thus we conclude that for these test conditions stable B100 will lead to stable B5 blends. We may also reasonably conclude that stable B100 will eliminate the most unstable B20 as measured by acid value, but the test cannot differentiate between intermediate and highly stable samples for acid number or sediment formation under these worst-case test conditions. This may be because these test conditions are too severe for B20 blends, such that stable and unstable blends are degraded. Additionally, because we cannot correlate results on this test to what actually happens in a vehicle fuel tank, it is not clear that the levels of insolubles shown in Figure 39, or the levels of acidity developed for the more stable B100 in Figure 37, actually indicate that these B20 blends would cause problems.

Blends were aged for 1 week at 80°C, then tested for thermal stability with D6468 (150°C for 180 minutes) to simulate reaction of the fuel in the high-temperature environment of the fuel injection system. Filter reflectance was measured and insolubles were quantified gravimetrically. For B5 blends the only sample exhibiting significant deposits was produced from a B100 with an induction time of 0.5 hours (see Figure 27). As shown in Figure 40, none of the B20 samples exhibited thermal instability based on reflectance. Most samples with B100 induction times of roughly 4 hours or less exhibited higher gravimetric deposits than other samples. We conclude that for B5 blends a stable B100 will lead to a thermally stable B5 blend, based on these test conditions, even after aging. For B20 blends additional data may be required, or the test conditions may be so severe that we cannot distinguish between stable and unstable samples. And as before, because we have no correlation of this test sequence (1 week at 80°C followed by D6468 thermal stability testing) with performance in actual engine fuel systems, we cannot be certain that any of the fuels tested would actually cause engine operational problems. This testing approach does not provide a clear performance delineation between fuels that are stable and those that are not under these conditions.

An important caveat is that these tests have all been conducted in glassware. We recommend that additional tests be performed in real equipment to validate these conclusions. For storage this might be accomplished by storing the fuels in drums at ambient conditions; for the fuel tank simulation perhaps blends from biodiesel of varying stability levels could be examined in stand-alone vehicle fuel systems, similar to the approach taken by Tsuchiya and coworkers.15 Unfortunately, there is no standard engine test for assessing the impact of fuel stability on fuel system and injector deposits and durability. Nevertheless, some testing with blends of varying stability in real fuels systems is required to validate these results.

How Do Diesel Fuel Properties Effect Stability?

Diesel fuel properties that might affect biodiesel blend stability include:

- Aromatic content. Higher aromatic fuels may be better able to hold polar biodiesel and diesel fuel oxidation products in solution.

- Sulfur content. Sulfur compounds can function as antioxidants.

- Diesel fuel oxidative and thermal stability.

Table 8 and Figure 15 show property and stability data for the diesel fuels used in this study. Total aromatic content ranges from about 8 to about 36 mass percent. One fuel (AL-27171) contains 340 ppm of sulfur; the remaining five contain less than 10 ppm. D2274 and D6468 indicate all six fuels are stable with little to distinguish one from another. Aging for 12 weeks in D4625 showed one fuel (AL-27175) beginning to show instability at 12 weeks. Notably this is the diesel fuel with the lowest aromatic content of 8%. Aging for 1 week at 80°C indicated little difference between the diesel fuels.

Examining accelerated test data, Figures 50 and 51 show how diesel fuel aromatic content affects D2274 total insolubles for B5 and B20 blends, respectively. In both cases the only samples exhibiting high levels of insolubles were prepared from B100 with induction times shorter than 1 hour, independent of fuel aromatic content. D4625 storage data for B5 blends show one sample with instability at 12 weeks (Figures 20–22, Appendix E). This fuel was prepared from an unstable B100 and the low-sulfur diesel fuel (AL-27171) with 36% aromatic content. This B5 blend was also unstable in simulated fuel tank aging and in high-temperature stability tests. A B20 blend prepared from these same blending components, as well as a second B20 prepared from an unstable biodiesel and a ULSD with 16% aromatic content also demonstrated instability on D4625.

To examine the issue of diesel fuel aromatic content in more detail, an unstable B100 (AL-27138) was blended with each diesel fuel at 20% and tested on the fuel tank aging and high-temperature stability simulation tests.

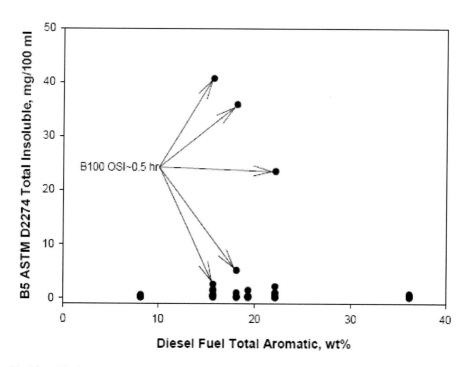

Figure 50. Diesel fuel aromatic content affect on D2274 total insolubles for B5 blends.

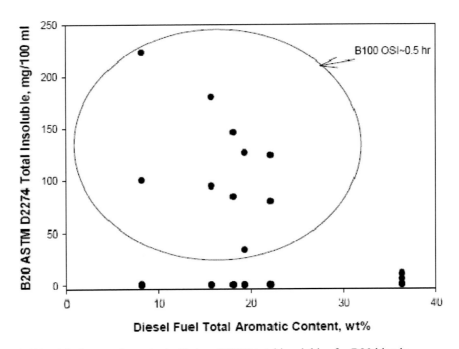

Figure 51. Diesel fuel aromatic content affect on D2274 total insolubles for B20 blends.

Results are shown in Figure 52. Four of the six samples produced total insolubles after 1 week of aging above 100 mg/100 ml, somewhat higher than observed when a range of B20 samples from various biodiesels were subjected to this test (Figure 39).

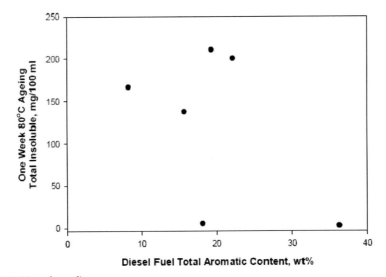

Figure 52. (Continued).

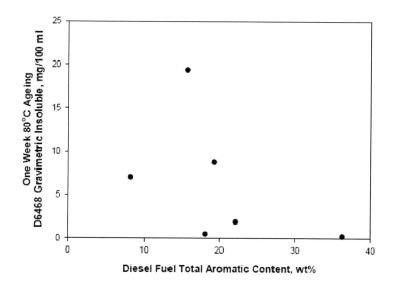

Figure 52. Results of 1 week at 80°C aging test followed by D6468 thermal stability test for B20 samples prepared from B100 AL-27138 and all diesel samples.

A blend of AL-27138 (B100) and AL-27151 (ULSD) was tested twice, once in November 2006 for the dataset shown in Figure 39 and again in April 2007 for the dataset shown in Figure 52. In November this sample produced 35.7 mg/100 ml; in April it generated 200 mg/100 ml, suggesting that the already unstable B100 had aged further in the intervening months. Nevertheless, these tests show no correlation between stability and diesel fuel aromatic content.

Our results suggest that diesel fuel aromatic content does not play a role in the stability of biodiesel blends and that B100 stability is the most important factor, consistent with the discussion in the previous section. However, we have noted the antagonistic effect where D2274 deposits for biodiesel blends are much higher than would be predicted based on the individual stability of the petroleum and biodiesel blend components. The cause of this antagonism is unknown, but based on these results does not appear to be related to aromatic content. The antagonism requires an unstable biodiesel, however, and perhaps unstable biodiesel degrades to produce polar components that are soluble in oxidized B100 but insoluble in any diesel fuel, independent of aromatic content.

The impact of diesel fuel sulfur content is less easily assessed because the dataset consists of essentially five samples with very low sulfur content (below 10 ppm) and one with 340 ppm.

The accelerated and real-world simulation results show that when an unstable B100 is blended with the 340 ppm sulfur diesel fuel an unstable blend is formed. Because there was little difference between the diesel fuels for stability, the impact of diesel stability on biodiesel blend stability cannot be assessed from these data. These results also support the idea that B100 stability is the main factor that affects the stability of B5 and B20 blends. However, an important caveat is that the diesel fuels included here were obtained before ULSD was mandated in the United States. Confirmatory testing should be conducted with fuels that are more representative of in-use fuels today.

How Do Antioxidant Additives Affect Stability?

The hindered phenolic antioxidants tested here prevented oxidative degradation of B100 in the storage simulation, and prevented degradation of biodiesel blends in the storage, fuel tank, and high-temperature simulations. The dramatic suppression in insoluble formation for the highly unstable B100 AL-27138 (Figure 14), and for B5 (Figure 23) and B20 (Figure 35) blends produced from it, during the 12-week storage test is striking, even for cases where the antioxidant did not increase B100 induction times to longer than 3 hours. For B5 blends, antioxidants also prevented insolubles from forming in the fuel tank and high-temperature simulations. In some B20 cases the additive treat rates were too low to prevent insolubles from forming for the fuel tank and high-temperature stability simulations, but did reduce insolubles formation. In these cases, a higher treat rate eliminated insolubles. No effort was made to optimize additive treat rate. Additionally, the amount of active ingredient in the additives is unknown. So the results mainly confirm that additives can be effective, but individual producers and blenders will have to select appropriate additives and determine optimal treat rates for their specific situations.

Is a 3-Hour B100 Induction Time Adequate?

Beginning with the publication of ASTM D6751-06b in January 2007, B100 is required to have a minimum 3-hour induction time. This requirement is largely based on the accelerated stability data and analysis presented in the section titled, Is B100 Stability a Predictor of Blend Stability? Additional data are now available from the aging simulation tests (storage, vehicle fuel tank, and high temperature) and the extent to which these new data support the 3-hour minimum induction time requirement is of interest.

Storage Stability
As discussed, none of the accelerated stability tests were particularly strong predictors of the stability of B100 on the D4625 test at 43°C. However, as indicated in Figure 43, having a minimum induction time of 3 hours limits acid value increase over 4 weeks of storage on this test to less than 0.2 mg KOH/g. Figure 11 shows that at 4 weeks the only samples that had gone out of specification for acid value had initial induction times of 3.1 hours or lower. Our experience agrees with that of other researchers,[5,7,8,9,14,15,17] that B100 samples with longer induction times are more stable in real-world settings. Thus, we take these results to mean that most biodiesel samples, regardless of initial induction time, will begin to oxidize immediately during storage. If induction time is near or below the 3-hour limit, the B100 will most likely go out of specification for either stability or acid value within 4 months. For longer induction times the biodiesel might be stored for several months. However, the results in Figure 9-12 indicate that even B100 with induction times longer than 7 hours will be out of specification for oxidation stability at only 4 months, although these samples may not have shown a significant increase in acidity or in deposit formation. The 3-hour limit appears to be adequate to prevent oxidative degradation for B5 and B20 blends in storage for up to 12 months (assuming that we accept the relationship that 1 week on D4625 is equivalent to 1 month of real-world storage at 21°C for blends).

Given these observations, we recommend that the 3-hour limit be considered as adequate to protect B100 and blends in storage, and that additional data on B100 storage be acquired at more realistic storage temperatures and for longer periods. Blenders of biodiesel and others

who need to store B100 should be made aware of limitations on the number of months this material can be stored. In developing a follow-up testing plan, the actual B100 storage conditions should be considered. In many cases, B100 may be stored in aboveground tanks that are heated to roughly 15°C during the winter months, but might experience higher temperatures during the summer months.

Stability in Simulated Vehicle Fuel Tank and High-Temperature Environments

The results of B5 testing shown in Figures 24 to 27 clearly indicate that B100 with a 3-hour or longer induction time produced B5 blends that were stable on the testing protocols employed here. Even if these test conditions are later found to have been too severe, the B5 blends did not exhibit instability if the B100 had at least a 3-hour induction time.

For B20 blends the situation is less clear cut. For the most part there was little to distinguish blends produced from minimum 3-hour induction time B100 from blends produced from less stable biodiesel. As shown in Figure 38, the B20 induction time was a better predictor of acid formation on the fuel tank aging simulation, as B20 induction time of roughly 4 hours was adequate to prevent a large increase in acidity.

Compared to the fuel tank aging simulation test, the D2274 test is conducted at conditions of 95°C with bubbling of oxygen through the fuel for 16 hours, in some ways more severe than the 80°C for 1 week test conditions. Increased contact of B100 and air via mixing or bubbling has been shown to produce a dramatic increase in oxidation rates.7 Figure 48 shows that B100 with a minimum 3-hour induction time produces stable B20 on this test. The accelerated test results shown in Figure 48 strongly support the conclusion that the B100 minimum 3-hour induction time produces B20 that is stable in the vehicle fuel tank and high-temperature environments. In other ways the 1-week storage test conditions may be more severe. Temperature and air contacting are less severe than D2274, but the test lasts more than 10 times longer (168 hours versus 16 hours). Perhaps we could better discriminate between stable and unstable B20 by reducing the test time from 1 week to 2 or 3 days, especially given that the test temperature of 80°C is likely higher than would be encountered in a real vehicle fuel tank over extended periods without being fueled with fresh fuel. Also, because Figure 47 indicates that if B100 induction time is longer than 3 hours, B20 induction time is almost always longer than 6 hours; therefore, the idea of a separate stability requirement for B20 may not provide additional protection. So although additional data are needed to confirm, the picture emerging from the available data is that B20 blends prepared from B100 with a minimum 3-hour induction time are stable in vehicle fuel tank and high-temperature engine fuel system environments.

CONCLUSIONS AND RECOMMENDATIONS

The B100 samples examined here show a broad distribution of stability on accelerated tests, with induction time (OSI) results ranging from less than 1 hour to more than 9 hours and ASTM D2274 total insolubles ranging from less than 2 mg/100 ml to nearly 18 mg/100 ml. Samples that produce very high levels of iso-octane insolubles on the D2274 test also produced very high levels of total insolubles. This suggests that little new information is acquired by measuring iso-octane insolubles. The change in acid value of the filtered liquid over the D2274 test was closely correlated with total insolubles and the initial induction time. The accelerated test data indicate that if the B100 stability is above a roughly 3-hour induction time, blends prepared from that B100 appear to be stable on the OSI and D2274 tests.

The D4625 long-term storage results for B100 indicate that most biodiesel samples, regardless of initial induction time, will begin to oxidize immediately during storage. If induction time is near or below the 3-hour limit, the B100 will most likely go out of specification for either stability or acid value within 4 months. Even B100 with induction times longer than 7 hours will be out of specification for oxidation stability at only 4 months, although these samples may not have shown a significant increase in acidity or in deposit formation. B100 should probably not be stored longer than several months in any case unless it is treated with a synthetic antioxidant. Blenders of biodiesel and others who need to store B100 should be made aware of limitations on the number of months this material can be stored.

The 3-hour B100 induction time limit appears to be adequate to prevent oxidative degradation for both B5 and B20 blends in storage for up to 12 months. Questions remain as to whether 1 week of storage on the D4625 test for biodiesel and biodiesel blends equates to 1 month of real world storage, as is generally assumed for petroleum-derived fuels.

For tests that simulated fuel tank aging and high-temperature stability, we conclude that stable B100 (longer than 3 hour OSI) leads to stable B5 blends. For B20, considering the entire dataset available today, we provisionally conclude that stable B100 (longer than 3 hour OSI) leads to stable B20 blends, but the test cannot differentiate between intermediate and highly stable samples for acid number or sediment formation under these worst-case test conditions. Anecdotal and detailed field data collections on B20 do not show a propensity for significant vehicle fuel filter clogging. Additional work is required to confirm this or to determine whether an additional stability test is required for the B20 blend.

These results support the idea that B100 stability is the main factor that affects the stability of B5 and B20 blends, independent of diesel fuel aromatic content, sulfur level, or stability. However, the diesel fuels included here were obtained before ULSD was mandated in the United States. Confirmatory testing should be conducted with fuels that are more representative of actual in-use fuels. Additionally, the factors leading to the negative antagonism for deposit formation that occurs with blending of unstable B100 in to diesel fuel remain unknown, and future studies should endeavor to understand this phenomenon.

The hindered phenolic antioxidants tested here prevented oxidative degradation of B100 in the storage simulation, and prevented degradation of biodiesel blends in the storage, fuel tank, and high-temperature simulations. Individual producers, blenders, and additive suppliers will have to select appropriate additives and determine optimal treat rates for their specific situations.

We recommend that additional data be acquired in real equipment to verify these conclusions. For storage this might be accomplished with storage in drums at ambient conditions; for the fuel tank simulation perhaps blends from biodiesel of varying stability levels could be examined in stand-alone vehicle fuel systems or more intense monitoring and measurement of fuel in actual use. Unfortunately there is no standard engine test to assess the impact of fuel stability on fuel system and injector deposits and durability. Nevertheless, some testing with blends of varying stability in real fuel systems is required to confirm these results.

REFERENCES

[1] Frankel, E.N. 2005. Lipid Oxidation, 2nd Edition. The Oily Press, Ltd, Bridgewater.

[2] Waynick, J.A. 2005. Characterization of Biodiesel Oxidation and Oxidation Products: CRC Project No. AVFL-2b. National Renewable Energy Laboratory, NREL/TP-540-39096.

[3] Cosgrove, J.P., Church, D.F., Pryor, W.A. 1987. "The Kinetics of the Autooxidation of Polyunsaturated Fatty Acids." Lipids 22:299–304.

[4] Knothe, G. 2002. "Structure Indices in FA Chemistry." J. Am. Oil Chem. Soc. 79:847–854.

[5] Westbrook, S.R. 2005. An Evaluation and Comparison of Test Methods To Measure the Oxidation Stability of Neat Biodiesel. Subcontract Report. National Renewable Energy Laboratory, NREL/SR-540-38983.

[6] McCormick, R.L., Ratcliff, M., Moens, L., Lawrence, R. 2007. "Several Factors Affecting the Stability of Biodiesel in Standard Accelerated Tests." Fuel Processing Technology, 88:651-657.

[7] Bondioli, P., Gasparoli, A., Della Bella, L., Tagliabue, S., Toso, G. 2003. "Biodiesel Stability under Commercial Storage Conditions over One Year." Eur. J. Lipid Sci. Technol. 105:735–741.

[8] Bondioli, P., Gasparoli, A., Della Bella, L., Tagliabue, S. 2002. "Evaluation of Biodiesel Storage Stability Using Reference Methods." Eur. J. Lipid Sci. Technol. 104:777–784.

[9] Mittelbach, M., Gangl, S. 2001. "Long Storage Stability of Biodiesel Made from Rapeseed and Used Frying Oil. J. Am. Oil Chem. Soc. 78:573–577.

[10] Fang, H., McCormick, R.L., 2006. "Spectroscopic Study of Biodiesel Degradation Pathways." SAE Techn. Pap. No. 2006-01-3300.

[11] Tatandjiiska, R.B., Marekov, I.N., NikilovaDamyanova, B.M. 1996. "Determination of Triacylglycerol Classes and Molecular Species in Seed Oils with High Content of Linoleic And Linolenic Acids." J. Sci. Food. Agric. 72:403–410.

[12] Van Gerpen, J.H., Hammond, E.G., Yu, L., Monyem, A., 1997. "Determining the Influence of Contaminants on Biodiesel Properties." SAE Techn. Pap. No. 971685.

[13] Terry, B., McCormick, R.L., Natarajan, M. 2006. "Impact of Biodiesel Blends on Fuel System Component Durability." SAE Techn. Pap. No. 2006-01-3279.

[14] Blassnegger, J. 2003. "Biodiesel as an Automotive Fuel: Bench Tests." Stability of Biodiesel – Used as a Fuel for Diesel Engines and Heating Systems. Presentation of the BIOSTAB project results. Proceedings. Graz, July 3, 2002. BLT Wieselburg, Austria.

[15] Tsuchiya, T., Shiotani, H., Goto, S., Sugiyama, G., Maeda, A. 2006. "Japanese Standard for Diesel Fuel Containing 5% of FAME; Investigation on Oxidation Stability of FAME Blended Diesel." SAE Techn. Pap. No. 2006-01-3303.

[16] Stavinoha, L.L., Howell, S. 1999. Potential Analytical Methods for Stability Testing of Biodiesel and Biodiesel Blends. SAE Techn. Pap. No. 1999-01-3520.

[17] Miyata, I., Takei,Y., Tsurutani, K., Okada, M. 2004. "Effects of Bio-Fuels on Vehicle Performance: Degradation Mechanism Analysis of Bio-Fuels." SAE Techn. Pap. No. 2004-01-3031.

[18] Bondioli, P., Gasparoli, A., Della Bella, L., Tagliabue, S., Lacoste, F., Lagardere, L. 2004. The Prediction of Biodiesel Storage Stability. Proposal for a Quick Test. Eur. J. Lipid Sci. Technol. 106:822-830.

[19] McCormick, R.L., Alleman, T.L., Ratcliff, M., Moens, L., Lawrence, R. 2005. "Survey of the Quality of Biodiesel and Biodiesel Blends in the United States in 2004." National Renewable Energy Laboratory, NREL/TP-540-38836.

[20] Proc, K., Barnitt, R., Hayes, R.R., McCormick, R.L., Ha, L., Fang, H.L. 2006. "100,000 Mile Evaluation of Transit Buses Operated on Biodiesel Blends (B20)." SAE Techn. Pap. No. 2006-01-3253.

APPENDIX A: B100 CHARACTERIZATION AND STABILITY TEST DATA

Sample Identification	Method	AL-27128-F	AL-27129-F	AL-27137-F	AL-27138-F	AL-27140-F	AL-27141-F
Oxidation Stability, hours	EN 14112						
Replicate 1		4.22	3.09	6.41	0.13	0.17	5.74
Replicate 2		4.18	3.10	6.55	0.62	0.17	5.75
Replicate 3		4.24	3.14	6.61	0.59	0.13	5.11
Mean		4.21	3.11	6.52	0.45	0.16	5.53
Oxidation Stability, mg/ 100 ml	Modified ASTM D2274						
Replicate 1							
Filterable insolubles		4.7	1.3	0.7	6.1	7.6	0.2
Adherent Insolubles		1.6	1.0	1.1	0.9	2.1	0.1
Total Insolubles		6.3	2.3	1.8	7.0	9.7	0.3
Iso-Octane Insolubles		197.4	4.7	2.4	4.0	64.1	0.7
Replicate 2							
Filterable insolubles		5.0	1.0	0.2	6.9	10.5	0.2
Adherent Insolubles		1.6	0.5	0.0	1.3	2.0	0.1
Total Insolubles		6.6	1.5	0.2	8.2	12.5	0.3
Iso-Octane Insolubles		198.0	0.4	2.8	4.8	73.0	0.5
Mean Total Insolubles		6.5	1.9	1.0	7.6	11.1	0.3

Appendix A: (Continued).

	Method						
Mean Iso-Octane Insolubles		197.7	2.6	2.6	4.4	68.6	0.6
Oxidation Stability, Pressure Vessel	ASTM D525						
Induction Period Method, minutes		341	169	34	80	71.0	335.0
Thermal Stability, deposit, mg	ASTM D6468 Modified	0.3	0.1	0.3	0.3	0.4	0.2
Total Acid Number, mg KOH/g	ASTM D664	0.23	0.41	0.05	0.33	0.2	0.13
Total Acid Number after D2274, mg KOH/g							
Replicate 1		1.84	2.92	0.24	2.75	3.83	0.79
Replicate 2		2.03	2.07	0.15	3.26	4.22	0.64
Mean		1.94	2.50	0.20	3.01	4.03	0.72
Total Glycerin, %(mass)	ASTM D6584	0.103	0.081	0.144	0.016	0.022	0.121
Free Glycerin, %(mass)	ASTM D6584	0.009	<0.001	0.002	0.002	0.015	0.005
Oxidation Stability, hours	EN 14112						
Replicate 1		4.56	3.69	3.70	9.28	7.82	7.24
Replicate 2		4.46	4.02	4.01	9.33	7.76	7.24
Replicate 3		4.40	4.10	4.16	9.16	7.69	7.31
Mean		4.47	3.94	3.96	9.26	7.76	7.26
Oxidation Stability, mg/100 ml	Modified ASTM D2274						
Replicate 1							
Filterable insolubles		0.9	2.0	2.0	0.3	0.8	0.4

Sample Identification	Method	AL-27128-F	AL-27129-F	AL-27137-F	AL-27138-F	AL-27140-F	AL-27141-F
Adherent Insolubles		0.2	0.3	0.4	0.1	0.1	0.1
Total Insolubles		1.1	2.3	2.4	0.4	0.9	0.5
Iso-Octane Insolubles		0.3	1.4	2.7	0.4	22.1	6.2
Replicate 2							
Filterable insolubles		0.9	2.2	2.0	0.2	0.7	0.3
Adherent Insolubles		0.2	0.1	0.2	0.0	0.1	0.2
Total Insolubles		1.1	2.3	2.2	0.2	0.8	0.5
Iso-Octane Insolubles		0.7	1.3	0.7	0.4	15.7	6.5
Mean Total Insolubles		1.1	2.3	2.3	0.3	0.9	0.5
Mean Iso-Octane Insolubles		0.5	1.4	1.7	0.4	18.9	6.4
Oxidation Stability, Pressure Vessel	ASTM D525						
Induction Period Method, minutes		507	62	721	649	325	418
Thermal Stability, deposit, mg	ASTM 6468 Modified	0.3	0.4	0.4	0.5	0.4	0.1
Total Acid Number, mg KOH/g	ASTM D664	0.07	0.39	0.49	0.08	0.69	0.09
Total Acid Number after D2274, mg KOH/g							
Replicate 1		1.11	1.91	2.18	0.19	1.56	0.73
Replicate 2		0.90	1.83	2.13	0.20	1.37	0.68

Appendix A: (Continued).

		1.01	1.87	2.16	0.20	1.47	0.71
Total Glycerin, %(mass)	ASTM D6584	0.216	0.221	0.192	0.161	0.121	0.15
Free Glycerin, %(mass)	ASTM D6584	0.003	0.004	0.005	<0.001	<0.001	0.001
Oxidation Stability, hours	EN 14112						
Replicate 1		9.36	0.14	2.60	4.67	7.18	0.18
Replicate 2		9.36	0.18	2.59	4.56	7.26	0.14
Replicate 3		9.35	0.18	2.27	4.47	7.52	0.18
Mean		9.36	0.17	2.49	4.57	7.32	0.17
Oxidation Stability, mg/ 100 ml	Modified ASTM D2274						
Replicate 1							
Filterable insolubles		0.2	12.3	9.4	1.5	0.2	11.0
Adherent Insolubles		0.1	1.3	1.5	0.5	0.1	2.1
Total Insolubles		0.3	13.6	10.9	2.0	0.3	13.1
Iso-Octane Insolubles		0.2	87.6	94.6	2.8	0.5	120.9
Replicate 2							
Filterable insolubles		0.2	11.1	8.5	1.5	0.2	18.0
Adherent Insolubles		0.0	1.5	1.3	0.8	0.1	4.1
Total Insolubles		0.2	12.6	9.8	2.3	0.3	22.1
Iso-Octane Insolubles		0.2	ND	95.8	2.8	0.0	126.7

Appendix A: (Continued).

Sample Identification	Method	AL-27128-F	AL-27129-F	AL-27137-F	AL-27138-F	AL-27140-F	AL-27141-F
Mean Total Insolubles		0.3	13.1	10.4	2.2	0.3	17.6
Mean Iso-Octane Insolubles		0.2	87.6	95.2	2.8	0.3	123.8
Oxidation Stability, Pressure Vessel	ASTM D525						
Induction Period Method, minutes		No break	150	270	487	512	159
Thermal Stability, deposit, mg	ASTM D6468 Modified	0.5	0.1	0.3	0.3	0.6	0.5
Total Acid Number, mg KOH/g	ASTM D664	0.08	1.31	0.29	0.11	0.51	0.46
Total Acid Number after D2274, mg KOH/g							
Replicate 1		0.21	5.54	3.21	1.05	0.70	5.12
Replicate 2		0.27	ND	2.93	1.19	0.69	6.40
Mean		0.24	5.54	3.07	1.12	0.70	5.76
Total Glycerin, %(mass)	ASTM D6584	0.15	0.132	0.298	0.225	0.158	0.188

Appendix A: (Continued).

Sample Identification	Method	AL-27161-F
Oxidation Stability, hours	EN 14112	
Replicate 1		5.13
Replicate 2		5.11
Replicate 3		5.09
Mean		5.11
Oxidation Stability, mg/ 100 ml	Modified ASTM D2274	
Replicate 1		
Filterable insolubles		1.0
Adherent Insolubles		0.1
Total Insolubles		1.1
Iso-Octane Insolubles		0.60
Replicate 2		
Filterable insolubles		0.8
Adherent Insolubles		0.2
Total Insolubles		1.0
Iso-Octane Insolubles		0.80
Mean Total Insolubles		1.1
Mean Iso-Octane Insolubles		0.7
Oxidation Stability, Pressure Vessel	ASTM D525	
Induction Period Method, minutes		No break
Thermal Stability, deposit, mg	ASTM D6468 Modified	0.2
Total Acid Number, mg KOH/g	ASTM D664	0.37
Total Acid Number after D2274, mg KOH/g		
Replicate 1		1.63
Replicate 2		1.50
Mean		1.57
Total Glycerin, %(mass)	ASTM D6584	0.151
Free Glycerin, %(mass)	ASTM D6584	0.003

Weeks	Property	Method	Units	AL-27128	AL-27129	AL-27137	AL-7138	AL-27141	AL-27148	AL-27152	AL-27160
0 Weeks	IsoOctane Insoluble	D 4625, mod	mg/100 ml	3.9	0.1	1.9		2.3	0.1	0.1	0.5
	Peroxide Value	D 3703		217	105	12		98	8	43	17
	Viscosity @ 40°C	D 445		4.45	5.12	4.10		4.09	4.67	4.47	4.86
	Polymer Content	ISO 16931		2.63	0.17	0.82		1.03	1.46	2.17	4.91
	Total Acid No.	D 664	mg KOH/g	0.23	0.41	0.05	0.33	0.13	0.69	0.09	0.46
	Total insoluble	D2274	mg/100ml	6.45	1.9	1.015	7.6	0.3	0.9	0.5	17.6
	IsoOctane Insols	D2274	mg/100ml	197.7	2.6	2.6	4.4	0.6	18.9	6.4	123.8
	Rancimat IP		h	4.21	3.11	6.52	0.45	5.53	7.76	7.26	0.17
4 Weeks	Filterable Insols	D 4625, mod	mg/100 ml	0.7	0.1	1.03	1.8	0.4	0.9	0.4	0.7
	Adherent Insols	D 4625, mod	mg/100 ml	0	0	0	0.1	1	0	0.1	0.2
	Total Insols	D 4625, mod	mg/100 ml	0.7	0	1.3	1.9	1.4	0.9	0.5	0.9
	IsoOctane Insols	D 4625, mod		14.8	0.7	3.3	0.2	4	1.2	3.8	10.3
	Peroxide Value	D 3703		601	633	403		841	65	721	566
	Viscosity @ 40°C	D 445		5.5	4.78	4.13		4.22	4.69		5.08
	Polymer Content	ISO 16931		2.8	0.49	0.89		1.26	1.5	2.29	5.63
	Total Acid No.	D 664	mg KOH/g	0.41	0.53	0.07	0.57	0.19	0.74	0.24	0.74
	Rancimat IP		h	0.88	0.17	1.91		1.67	1.23	2.62	0.54
8 Weeks	Filterable Insols	D 4625, mod	mg/100 ml	1	0.1	0.3	1.1	0.3	0.2	1.1	2.2
	Adherent Insols	D 4625, mod	mg/100 ml	0.1	0.1	0.1	0.1	0.1	0.1	0.1	0.1
	Total Insols	D 4625, mod	mg/100 ml	1.1	0.2	0.4	1.2	0.4	0.3	1.2	2.3
	IsoOctane Insols	D 4625, mod		59.7	0.5	3.6	1.9	3.6	1.9	28.5	55.3
	Peroxide Value	D 3703		1237	1759	2039		1654	225	903	1000

Appendix B: (Continued).

Viscosity @ 40°C	D 445		5.75	5.03	4.54		4.49	4.73	4.81	5.36
Polymer Content	ISO 16931		3.59	1.33	1.92		2.01	1.54	2.82	7.2
Total Acid No.	D 664	mg KOH/g	1.22	1.41	0.98		0.98	1.64	1.22	1.76
Rancimat IP		h	>12	>12	>12	12	0.4	1.29	0.63	0.32
12 Weeks Filterable Insols	D 4625, mod	mg/100 ml	3.1	1.4	1	5 9	4.1	3.2	4.2	2.8
Adherent Insols	D 4625, mod	mg/100 ml	0.7	0.5	0.7	0 3	0.4	0.2	0.3	0.1
Total Insols	D 4625, mod	mg/100 ml	3.8	1.9	1.7	6 2	4.5	3.4	4.5	2.9
IsoOctane Insols	D 4625, mod		17.6	1.1	7.3		6.4	13.1	74.7	97.4
Peroxide Value	D 3703		2037	2026	3099		2951	881	1762	1208
Viscosity @ 40°C	D 445		4.83	4.93	4.72		4.67	4.84	5.04	5.53
Polymer Content	ISO 16931		4.45	1.99	3.07		3.26	2.02	4.18	8.42
Total Acid No.	D 664	mg KOH/g	1.10	1.10	0.69	3 18	0.68	1.04	1.32	1.77
Rancimat IP		h	>12	>12	>12		0.58	0.98	>12	>12

B100 Storage with Antioxidants			AL-27138 Neat	AL-27138 + 1000 ppm A	AL-27138 + 2000 ppm A	AL-27138 + 1000 ppm B	AL-27138 + 2000 ppm B	AL-27129 + 1000 ppm A	AL-27129 + 1000 ppm B
				After 4 Weeks Storage at 43°C					
Filterable Insolubles	D 4625, mod	mg/100 ml	1.8	0.2	0.2	0.2	0.2	0.3	0.1
Adherent Insolubles	D 4625, mod	mg/100 ml	0.1	0.1	0.1	0.1	0.1	0.1	0.1
Total Insolubles	D 4625, mod	mg/100 ml	1.9	0.3	0.3	0.5	0.3	0.4	0.2
Isooctane Insoluble	D 4625, mod	mg/100 ml	0.2	0.4	0.7	0.4	0.2	0.2	0.2
Total Acid Number	D 664	mg KOH/g	0.57	0.35	0.35	0.32	0.35	0.43	0.43
After 8 Weeks Storage at 43°C									
Filterable Insolubles	D 4625, mod	mg/100 ml	1.1	0.3	0.3	0.5	0.3	0.4	0.4

	Method	Units						
Adherent Insolubles	D 4625, mod	mg/100 ml	0.1	0.7	0.2	0.1	0.2	0.1	0.1
Total Insolubles	D 4625, mod	mg/100 ml	1.2	1.0	0.5	0.4	0.5	0.5	0.5
Isooctane Insoluble	D 4625, mod	mg/100 ml	1.9	1.6	1.8	1.6	5.8	0.5	0.3
Total Acid Number	D 664	mg KOH/g	1.20	*	1.28	1.29	1.23	0.42	0.40
After 12 Weeks Storage at 43°C									
Filterable Insolubles	D 4625, mod	mg/100 ml	5.9	0.3	0.4	0.3	0.4	0.1	0.1
Adherent Insolubles	D 4625, mod	mg/100 ml	0.3	0.1	0.2	0.2	0.4	0.1	0
Total Insolubles	D 4625, mod	mg/100 ml	6.2	0.4	0.6	0.5	0.8	0.2	0.1
Isooctane Insoluble	D 4625, mod	mg/100 ml	1.4	3.2		1.5	2.1	0.7	0.6
Total Acid Number	D 664	mg KOH/g	3.18	0.41	0.37	0.66	0.41	0.42	0.48

APPENDIX C: D4625 RESULTS FOR PETROLEUM-DERIVED DIESEL FUELS

				AL-27150	AL-27151	AL-27166	AL-2717	AL-27175	AL-27176
4 Weeks	Filterable Insols	D 4625, mod	mg/100 ml	0.0	0.1	< 0.1	0.3	0.1	< 0.1
	Adherent Insols	D 4625, mod	mg/100 ml	0.1	0.3	< 0.1	0.1	< 0.1	0.0
	Total Insols	D 4625, mod	mg/100 ml	0.1	0.4	0.1	0.4	0.1	< 0.1
8 Weeks	Filterable Insols	D 4625, mod	mg/100 ml	0.0	0.1	0.1	0.3	0.1	0.1

Appendix C: (Continued).

	D 4625, mod	mg/100 ml						
Adherent Insols	D 4625, mod	mg/100 ml	0.0	0.2	0.0	0.4	0.0	0.1
Total Insols	D 4625, mod	mg/100 ml	0.0	0.3	0.1	0.7	0.2	0.3
Filterable Insols	D 4625, mod	mg/100 ml	0.0	0.1	0.1	0.2	2.6	0.3
Adherent Insols	D 4625, mod	mg/100 ml	0.3	0.6	0.2	0.6	0.3	0.3
12 Weeks — Total Insols	D 4625, mod	mg/100 ml	0.3	0.6	0.2	0.8	2.9	0.6

APPENDIX D: ACCELERATED TEST RESULTS FOR B5 BLENDS

CL#	Biodiesel	%	Diesel	Diesel IC BP, °	Rancimat IP, h	D2274, mod, mg/100 ml Filt Insols	D2274, mod, mg/100 ml Adh Insols	D2274, mod, mg/100 ml Tot Insols	D525* IP, h	D6468 Reflectance, %	D6468 mg/100ml
CL06-198	27128	95	27150	154	>20	0.8	0.6	1.4	No break	97.6	<0.1
CL06-199	27128	95	27151	167	>20	0.2	0.3	0.5	No break	99.0	0.2
CL06-200	27128	95	27166	165	>20	0.3	0.3	0.6	No break	97.8	<0.1
CL06-201	27128	95	27171	190	>20	0.5	0.2	0.7	No break	84.1	<0.1
CL06-202	27128	95	27175	164	>20	0.2	0.2	0.4	No break	93.8	<0.1
CL06-203	27128	95	27176	183	>20	0.2	0.1	0.3	No break	98.9	<0.1
CL06-210	27129	95	27150	154	>20	0.2	0.4	0.6	No break	95.9	<0.1
CL06-211	27129	95	27151	167	>20	0.8	0.2	1.0	No break	98.3	<0.1
CL06-212	27129	95	27166	165	>20	0.1	0.1	0.2	No break	98.7	<0.1
CL06-213	27129	95	27171	190	>20	0.1	0.1	0.2	No break	85.5	<0.1
CL06-	27129	95	27175	164	>20	0.5	0.1	0.6	No break	95.3	0.1

214											
CL06-215	27129	95	27176	183	>20	0.1	1.4	1.5	1.8	98.7	<0.1
CL06-223	27137	95	27150	154	>20	0.2	1.4	1.6	No break	97.0	<0.1
CL06-224	27137	95	27151	167	>20	0.1	0.9	1.0	No break	98.8	<0.1
CL06-225	27137	95	27166	165	>20	0.1	0.9	1.0	No break	98.6	0.1
CL06-226	27137	95	27171	190	>20	0.3	0.1	0.4	No break	91.5	<0.1
CL06-227	27137	95	27175	164	>20	0.2	0.1	0.3	No break	94.1	<0.1
CL06-228	27137	95	27176	183	18.5	0.3	0.1	0.4	No break	98.9	0.4
CL06-235	27138	95	27150	154	4.8	39	1.7	40.7	8.3	87.7	0.1
CL06-236	27138	95	27151	167	5.8	23	0.5	23.5	No break	98.3	<0.1
CL06-237	27138	95	27166	165	5.0	32	3.9	35.9	No break	90.3	0.3
CL06-238	27138	95	27171	190	>12	0.2	0.1	0.3	No break	32.1	0.3
CL06-239	27138	95	27175	164	6.9	0.1	0.1	0.2	No break	93	<0.1
CL06-240	27138	95	27176	183	0.7	0.2	0.1	0.3	No break	98.4	0.4
CL06-248	27141	95	27150	154	>12	0.1	0.5	0.6	No break	97.5	<0.1
CL06-249	27141	95	27151	167	>12	0.1	0.1	0.2	No break	99.0	<0.1
CL06-250	27141	95	27166	165	>12	0.2	0.1	0.3	No break	96.4	<0.1
CL06-251	27141	95	27171	190	>12	0.1	0.1	0.2	No break	87.7	<0.1

Appendix D: (Continued).

CL06-252	27141	95	27175	164	>12	0.2	0.1	0.3	No break	95.6	<0.1
CL06-253	27141	95	27176	183	>12	0.2	0.1	0.3	No break	99.3	<0.1
CL06-260	27148	95	27150	154	>12	0.2	0.1	0.3	No break	89.6	0.1
CL06-261	27148	95	27151	167	>12	0.1	0.1	0.2	No break	91.4	0.1
CL06-262	27148	95	27166	165	>12	0.2	0.1	0.3	No break	91.0	0.6
CL06-263	27148	95	27171	190	>12	0.1	0.3	0.4	No break	84.4	0.4
CL06-264	27148	95	27175	164	>12	0.2	0.2	0.4	No break	93.4	0.2
CL06-265	27148	95	27176	183	>12	0.1	0.1	0.2	No break	87.6	0.4
CL06-272	27152	95	27150	154	>12	0.1	0.1	0.2	No break	97.9	<0.1
CL06-273	27152	95	27151	167	>12	0.1	0.2	0.3	No break	99.1	<0.1
CL06-274	27152	95	27166	165	>12	0.1	0.1	0.2	No break	99.0	0.1
CL06-275	27152	95	27171	190	>12	0.1	0.3	0.4	No break	91.7	0.1
CL06-276	27152	95	27175	164	>12	0.1	0.2	0.3	No break	96.0	<0.1
CL06-277	27152	95	27176	183	>12	0.1	0.1	0.2	No break	99.2	0.1
CL06-284	27160	95	27150	154	>12	2.4	0.2	2.6	No break	48.6	2.0
CL06-285	27160	95	27151	167	>12	2.0	0.2	2.2	No break	72.0	0.5
CL06-286	27160	95	27166	165	>12	4.6	0.6	5.2	No break	65.1	0.7
CL06-287	27160	95	27171	190	5.5	0.6	0.2	0.8	No break	19.3	2.0
CL06-288	27160	95	27175	164	>12	0.1	0.1	0.2	No break	55.7	0.5
CL06-289	27160	95	27176	183	>12	0.2	0.1	0.3	No break	68.1	1.6

	Property	Method	Units	C L06-46 2 5% AL-7128 in 95% AL-27166	C L06-46 3 5% AL-7129 in 95% AL-27176	C L06-46 4 5% AL-7137 in 95% AL-27175	C L06-46 5 5% AL-7138 in 95% AL-27150	C L06-46 6 5% AL-7141 in 95% AL-27175	C L06-46 7 5% AL-7148 in 95% AL-27175	C L06-46 8 5% AL-7152 in 95% AL-27151	C L06-46 9 5% AL-7160 in 95% AL-27171
0 Weeks	Peroxide Value	D 3703		41.24	8.14	18.79	21.16	32.27	18.75	21.56	4.17
	Viscosity@40 C	D 445		1.72	1.61	2.37	1.67	2.31	2.34	2.36	3.04
	Total Acid No.	D 664	mg KOH/g	0.03	0.06	0.03	0.05	0.04	0.06	0.05	0.05
4 Weeks	Filterable Insols	D 4625, mod	mg/100 ml	0.2	0.3	0.1	-0.2	0.2	0.1	0	0.5
	Adherent Insols	D 4625, mod	mg/100 ml	1	0.6	0.2	0.5	0.2	-0.4	1	0.2
	Total Insols	D 4625, mod	mg/100 ml	1.2	0.9	0.3	0.3	0.4	-0.3	1	0.7
	Peroxide Value	D 3703		64.71	77.97	61.58	217.33	68.75	168.66	33.53	187.67
	Viscosity@40 C	D 445		1.76	1.6	2.33	1.66	2.31	2.34	2.38	3.01
	Total Acid No.	D 664	mg KOH/g	0.01	0.01	0.04	0	0.04	0.02	0.07	0.04
8 Weeks	Filterable Insols	D 4625, mod	mg/100 ml	0.3	0.5	0.2	0.5	0.3	0.3	0.2	0.9
	Adherent Insols	D 4625, mod	mg/100 ml	0.6	0.1	0	0.6	0.8	0.3	0.5	0.8
	Total Insols	D 4625, mod	mg/100 ml	0.9	0.6	0.2	1.1	1.1	0.6	0.7	1.7
	Peroxide Value	D 3703		17.58	10.38	12.39	44.71	23.92	4.79	13.98	10.8

Appendix E. (Continued).

	Property	Method	Units	C L06-46 2 5% AL-7128 in 95% AL-27166	C L06-46 3 5% AL-7129 in 95% AL-27176	C L06-46 4 5% AL-7137 in 95% AL-27175	C L06-46 5 5% AL-7138 in 95% AL-27150	C L06-46 6 5% AL-7141 in 95% AL-27175	C L06-46 7 5% AL-7148 in 95% AL-27175	C L06-46 8 5% AL-7152 in 95% AL-27151	C L06-469 5% AL-7160 in 95% AL-27171
	Viscosity@40 C	D 445		1.73	1.61	2.37	1.67	2.33	2.35	2.36	3.07
	Total Acid No.	D 664	mg KOH/g	0.01	0.03	0.01	0.02	0.01	0.02	0.07	0.04
12 Weeks	Filterable Insols	D 4625, mod	mg/100 ml	0.3	1.4	0.2	24.3	0.3	0.2	0.2	0.7
	Adherent Insols	D 4625, mod	mg/100 ml	1.3	0.7	1.3	22.7	1	0.5	0.6	0.8
	Total Insols	D 4625, mod	mg/100 ml	1.6	2.1	1.5	57	1.3	0.7	0.8	1.5
	Peroxide Value	D 3703		21.98	10.79	12.79	1001.99	25.18	4.80	20.37	9.59
	Viscosity@40 C	D 445		1.73	1.58	2.31	.69	2.31	2.36	2.34	2.98
	Total Acid	D 664	mg KOH/g	0.03	0.01	0.02	.32	0.04	0.06	0.05	0.06
1 Week	Filterable Insols	D 4625, mod @ 80C	mg/100 ml	0.4	0.4	1.3	8.9	1.3	3.2	3	1.2
	Adherent Insols	D 4625, mod @ 80C	mg/100 ml	0.9	0.4	0.6	149.5	0.9	0.6	0.7	0.7
	Total Insols	D 4625, mod @ 80C	mg/100 ml	1.3	0.8	1.9	248.4	2.2	3.8	3.7	1.9
	Peroxide Value	D 3703		45.99	42.38	41.56	3046.12	118.73	16.37	27.92	465.02
	Viscosity@40 C	D 445		2.31	1.85	3.73	2.71	3.69	3.6	2.94	3.6
	Gravimetric	D 6468	% Reflectance	98.8	99.1	91.8	77.3	91.1	74.1	97.6	52.4
			Heptane Rinse mg/150ml	0.15	0.3	0.15	75	0.15	0.3	0.6	0.45
	Total Acid No.	D 664	mg KOH/g	0.05	0.07	0.03	1.84	0.05	0.08	0.06	0.06

Antioxidant Treated B5 Blends

				CL06-331	CL06-332	CL06-333	CL06-334	CL06-516	CL06-518	CL06-520	CL06-522
				5% 27138 + 2000 ppm A, 95% 27151	5% 27138 + 1000 ppm B, 95% 27151	5% 27138 + 2000 ppm B, 95% 27151	5% 27129 + 1000 ppm A, 95% 27151	5% 27138 2000ppm A, 95% 27175	5% 27138 +1000ppm B, 95% 27175	5% 27138 2000ppm B, 95% 27175	5% 27138 1000ppm A, 95% 27175
0 Weeks	Filterable Insols	D 2274, mod	mg/100 ml	0.1	0	0	0	0.0	0.0	0.2	0.1
	Adherent Insols	D 2274, mod	mg/100 ml	0.2	0	0	0	0.2	0.0	0.1	0.1
	Total Insols	D 2274, mod	mg/100 ml	0.3	0	0	0	0.2	0.0	0.3	0.2
	Rancimat IP		h	>12	>12	>12	>12	>12	>12	>12	>12
4 Weeks	Filterable Insols	D 4625, mod	mg/100 ml	0.2	0.2	0.2	0.2	0.1	0.1	0.0	0.1
	Adherent Insols	D 4625, mod	mg/100 ml	0.5	0.5	0.6	0.3	0.2	0.5	0.0	0.0
	Total Insols	D 4625, mod	mg/100 ml	0.7	0.7	0.8	0.5	0.3	0.6	0.0	0.1
	Rancimat IP		h	7.75	9.57	10.46	0	>12	>12	>12	>12
	Total Acid No.	D 664	mg KOH/g	0.93	0.92	0.91	0.93				
8 Weeks	Filterable Insols	D 4625, mod	mg/100 ml	0.2	0.2	0.2	0.1	0.2	0.4	0.2	0.4
	Adherent Insols	D 462 5, mod	mg/100 ml	1.7	1.5	1.1	0.7	0.2	0.7	0.4	0.3
	Total Insols	D 4625, mod	mg/100 ml	1.9	1.7	1.3	0.8	0.4	1.1	0.6	0.7
	Rancimat IP		h	>12	9.87	>12	>12	>12	>12	>12	>12
	Total Acid No.	D 664	mg KOH/g	>12	9.91	>12	>12	>12	>12	>12	>12
12 Weeks	Filterable Insols	D 4625, mod	mg/100 ml	0.2	0.2	0.1	0.2	0.3	0.3	1.1	0.6

(Continued).

Antioxidant Treated B5 Blends				CL06-331	CL06-332	CL06-333	CL06-334	CL06-516	CL06-518	CL06-520	CL06-522
				5% 27138 + 2000 ppm A 95% 27151	5% 27138 + 1000 ppm B 95% 27151	5% 27138 + 2000 ppm B 95% 27151	5% 27129 + 1000 ppm A 95% 27151	5% 27138 2000pp m A, 95% 27175	5% 27138 +1000pp m B, 95% 27175	5% 27138 2000pp m B, 95% 27175	5% 27138 1000pp m A, 95% 27175
	Adherent Insols	D 4625, mod	mg/100 ml	0.5	0	-0.2	-0.2	1.2	0.7	0.7	0.3
	Total Insols	D 4625, mod	mg/100 ml	0.7	0.2	-0.1	0	1.5	1.0	1.8	0.9
	Rancimat IP		h	11.27	10.74	>12	>12				
	Total Acid No.	D 664	mg KOH/g	0.08	0.08	0.06	0.09				
1 Weeks	Filterable Insols	D 4625, mod	mg/100 ml	0.4	0.6	0.8	0.8	2.1	1.6	4.5	1.8
80C	Adherent Insols	D 4625, mod	mg/100 ml	0.3	0.4	0.1	0.3	0.5	-0.2	0.0	11.9
	Total Insols	D 4625, mod	mg/100 ml	0.7	1.0	0.9	11	2.6	1.4	4.5	13.7
	Thermal Stability	D 6468	% Reflectance	99.0	99.2	99.0	98.1	91.0	92.6	90.9	89.2
			Heptane Rinse mg/100ml	0.15	0.15	0.15	0.15	0.15	0.15	0.15	0.15

Additional Antioxidant Treated B5 Blends

			CL06-546	CL06-548	CL06-550	CL06-552
			5% 27138 +	5% 27138 +	5% 27129 +	5% 27129 +
			1000 ppm A	1000 ppm A	1000 ppm B	1000 ppm B
1 Weeks 80C			95% 27175	95% 27151	95% 27151	95% 27175
	Filterable Insols	D 4625, mod	mg/100 ml	21.5	89	2.5	13.4
	Adherent Insols	D 4625, mod	mg/100 ml	0.7	11	0.2	0.2
	Total Insols	D 4625, mod	mg/100 ml	22.2	100	2.7	13.6
	Thermal Stability	D 6468	% Reflectance	77.5	92	98	88
			Heptane Rinse mg/100ml	2.1	10.5	0.15	0.15

APPENDIX F: ACCELERATED TEST RESULTS FOR B20 BLENDS

CL#	Biodiesel	%	Diesel	Rancimat	D2274, mod, mg/100 ml			D525	D6468	
				IP, h	Filt Insols	Adh Insols	Tot Insols	IP, h	Reflectance, %	mg/100ml
CL06-192	27128	80	27150	8.9	0.1	0.2	0.3	No break	97.9	<0.1
CL06-193	27128	80	27151	10.2	0.1	0.1	0.2	No break	99.1	<0.1
CL06-194	27128	80	27166	8.2	0.2	0.5	0.7	No break	98.4	0.5
CL06-195	27128	80	27171	5.5	0.8	0.2	1	No break	91.0	<0.1
CL06-196	27128	80	27175	7.8	0.2	0.1	0.3	No break	94.7	<0.1
CL06-197	27128	80	27176	9.5	0.1	0.1	0.2	No break	98.9	<0.1
CL06-204	27129	80	27150	>12	0.3	0.1	0.4	11.5	97.6	<0.1

Appendix F. (Continued).

CL#	Biodiesel	%	Diesel	Rancimat IP, h	D2274, mod, mg/100 ml			D525 IP, h	D6468 Reflectance, %	mg/100ml
					Filt Insols	Adh Insols	Tot Insols			
CL06-205	27129	80	27151	>12	0.2	0.1	0.3	No break	98.4	<0.1
CL06-206	27129	80	27166	>12; 7.3	0.2	0.1	0.3	11.9	95.9	<0.1
CL06-207	27129	80	27171	>12	0.2	0.3	0.5	No break	89.4	1.2
CL06-208	27129	80	27175	>12	0.1	0.2	0.3	No break	97.9	<0.1
CL06-209	27129	80	27176	>12; 12	0.2	0.4	0.6	No break	99.0	<0.1
CL06-229	27137	80	27150	>12	0.2	0.4	0.6	No break	98.2	<0.1
CL06-230	27137	80	27151	>12	0.4	0.6	1	No break	98.8	<0.1
CL06-231	27137	80	27166	11.6	0.1	0.2	0.3	No break	99.0	0.2
CL06-232	27137	80	27171	>12	0.2	0.4	0.6	No break	82.1	0.9
CL06-233	27137	80	27175	>12	0.1	0.1	0.2	No break	98.0	<0.1
CL06-234	27137	80	27176	>12; 12	0.1	0.1	0.2	No break	98.9	<0.1
CL06-241	27138	80	27150	1.8	47.6	132.4	180	1.5	91.1	1.5
CL06-242	27138	80	27151	1.8	42.4	82.4	124.8	1.4	99.2	<0.1
CL06-243	27138	80	27166	1.9	36.8	109.4	146.2	2.2	91.3	0.2
CL06-244	27138	80	27171	3.5	6.1	4.8	10.9	No break	54.1	1.1
CL06-245	27138	80	27175	1.7	51.7	171.1	222.8	2.4	97.8	<0.1

CL#	Biodiesel	%	Diesel	Rancimat IP, h	D2274, mod, mg/100 ml			D525 IP, h	D6468 Reflectance, %	mg/100ml
					Filt Insols	Adh Insols	Tot Insols			
CL06-246	27138	80	27176	3.2	42.2	85	127.2	4.3	99.3	<0.1
CL06-254	27141	80	27150	12.2	0.2	0.1	0.3	No break	98.2	0.1
CL06-255	27141	80	27151	12.5	0.3	0.1	0.4	No break	99.3	0.1
CL06-256	27141	80	27166	10.6	0.3	0.1	0.4	No break	98.7	<0.1
CL06-257	27141	80	27171	>12	0.8	0.1	0.9	No break	94.2	<0.1
CL06-258	27141	80	27175	11.1	0.2	0.1	0.3	No break	98.8	<0.1
CL06-259	27141	80	27176	10.2	0.3	0.1	0.4	No break	99.3	0.3
CL06-266	27148	80	27150	>12	0.3	0.1	0.4	No break	97.3	<0.1
CL06-267	27148	80	27151	>12	0.2	0.1	0.3	No break	97.1	<0.1
CL06-268	27148	80	27166	>12	0.2	0.1	0.3	No break	97.1	0.1
CL06-269	27148	80	27171	>12	0.4	0.1	0.5	No break	92.6	<0.1
CL06-270	27148	80	27175	>12	0.4	0.1	0.5	No break	87.6	0.4
CL06-271	27148	80	27176	>12	0.5	0.1	0.6	No break	97.0	0.3
CL06-278	27152	80	27150	>12	0.5	0.1	0.6	No break	98.5	0.1
CL06-279	27152	80	27151	>12	0.2	0.1	0.3	No break	99.5	<0.1
CL06-280	27152	80	27166	>12	0.3	0.1	0.4	No break	99.0	0.2
CL06-281	27152	80	27171	>12	0.3	0.1	0.4	No break	96.0	0.4

Appendix F. (Continued).

CL#	Biodiesel	%	Diesel	Rancimat IP, h	D2274, mod, mg/100 ml			D525 IP, h	D6468	
					Filt Insols	Adh Insols	Tot Insols		Reflectance, %	mg/100ml
CL06-282	27152	80	27175	>12	1.1	0.1	1.2	No break	97.2	<0.1
CL06-283	27152	80	27176	>12	0.1	0.1	0.2	No break	99.4	0.5
CL06-290	27160	80	27150	2.3	34	60.9	94.9	3.9	93.7	0.1
CL06-291	27160	80	27151	2.7	38.8	41.6	80.4	3.3	90.3	<0.1
CL06-292	27160	80	27166	2.3	40.4	44.5	84.9	3.2	84.0	1.4
CL06-293	27160	80	27171	4.5	5.7	0.3	6	No break	90.5	0.8
CL06-294	27160	80	27175	2.9	48.1	52.2	100.3	No break	78.9	0.7
CL06-295	27160	80	27176	4.1	31.7	2	33.7	7.5	71.5	1.0

APPENDIX G: B20 BLEND STORAGE, FUEL TANK, AND HIGH-TEMPERATURE STABILITY RESULTS

			CL06-480	CL06-481	CL06-482	CL06-483	CL06-484	CL06-485	CL06-486	CL06-487
			20%AL-7128 in AL-27175-F	20%AL-7129 in AL-27150-F	20%AL-7137 in AL-27171-F	20%AL-7138 in AL-27151-F	20%AL-7141 in AL-27166-F	20%AL-7148 in AL-27166-F	20%AL-7152 in AL-27175-F	20%AL-7160 in AL-27176-F
0 Weeks	Peroxide Value	D 3703	108.51	28.52	66.07	2?4.84	75.37	61.45	74.30	46.91
	Viscosity@40C	D 445	2.56	1.89	3.10	?.51	1.92	1.94	2.56	1.91

	Parameter	Method	Units								
4 Weeks	Total Acid No.	D 664	mg KOH/g	0.06	0.09	0.01	0.09	0.02	0.17	0.02	0.08
	Filterable Insols	D 4625, mod	mg/100 ml	0.2	0.1	0.3	0.5	1.3	0.2	0.1	0.6
	Adherent Insols	D 4625, mod	mg/100 ml	<0.1	0	0.1	0.7	0.5	<0.1	<0.1	<0.1
	Total Insols	D 4625, mod	mg/100 ml	0.1	0.1	0.4	1.2	1.8	<0.1	0	0.3
	Peroxide Value	D 3703		55.56	26.36	17.19	1183.43	103.15	17.99	27.96	71.91
	Viscosity @ 40°C	D 445		2.53	1.93	3.14	2.62	1.94	1.96	2.56	1.88
8 Weeks	Total Acid No.	D 664	mg KOH/g	0.07	0.11	0.07	0.2	0.05	0.16	0.01	0.14
	Filterable Insols	D 4625, mod	mg/100 ml	0.2	0.1	0.2	64.8	0.3	0.3	0.3	4
	Adherent Insols	D 4625, mod	mg/100 ml	0.8	0.7	0.4	110.3	4.3	0.5	1.2	2.6
	Total Insols	D 4625, mod	mg/100 ml	1	0.8	0.6	175.1	4.6	0.8	1.5	6.6
	Peroxide Value	D 3703		99.11	43.96	82.73	2733.56	147.48	33.46	72.33	531.11
	Viscosity@40C	D 445		2.57	1.92	3.11	2.91	1.96	1.98	2.54	1.92
12 Weeks	Total Acid No.	D 664	mg KOH/g	0.09	0.11	0.33	1.7	0.08	0.19	0.06	0.35
	Filterable Insols	D 4625, mod	mg/100 ml	0.8	0.4	0.3	55.1	0.8	0.5	0.8	0.8
	Adherent Insols	D 4625, mod	mg/100 ml	0.3	2.6	0.4	141.5	1.2	0.1	0.4	0.4
	Total Insols	D 4625, mod	mg/100 ml	1.1	3.0	0.7	196.6	2.0	0.6	1.2	1.2
	Peroxide Value	D 3703		115.45	47.18	77.41	1379.57	162.24	30.39	93.49	1024.25

Appendix G. (Continued).

	Method	Units	CL06-480	CL06-481	CL06-482	CL06-483	CL06-484	CL06-485	CL06-486	CL06-487
			20%AL-7128 in AL-27175-F	20%AL-7129 in AL-27150-F	20%AL-7137 in AL-27171-F	20%AL-7138 in AL-27151-F	20%AL-7141 in AL-27166-F	20%AL-7148 in AL-27166-F	20%AL-7152 in AL-27175-F	20%AL-7160 in AL-27176-F
Viscosity @ 40°C	D 445		2.52	1.93	3.14	3.02	1.95	1.99	2.56	1.93
Total Acid No.	D 664	mg KOH/g	0.02	0.13	0.39	2.19	0.09	0.18	0.09	0.66
1 Weeks Filterable Insols	D 4625, mod @80C	mg/100 ml	56	17.7	10	26.1	25.4	48.3	0.8	15.3
Adherent Insols	D 4625, mod @80C	mg/100 ml	2.4	0.8	2	9.6	0.9	0.2	0.3	0.6
Total Insols	D 4625, mod @80C	mg/100 ml	58.4	18.5	12	35.7	26.3	48.5	1.1	15.9
Peroxide Value	D 3703		353.15	4474.52	494.89	2230.16	1336.25	1325.35	513.02	1167.71
Viscosity @ 40°C	D 445		5.38	5.16	4.09	5.00	3.66	3.54	4.99	3.93
Gravimetric	D 6468	% Reflectance	91.5	89.6	91.1	88.5	99.5	77.4	95.5	86.6
		Heptane Rinse	4.95	12.3	0.3	18.6	0.45	3.6	0.3	2.4
Total Acid No.	D 664	mg KOH/g	0.66	2.88	0.11	1.79	0.25	0.63	0.33	0.89

Antioxidant Treated B20 Blends

	Method	Units	CL06-335	CL06-336	CL06-337	CL06-338	CL06-517	CL06-519	CL06-521	CL06-523
			20% 27138 +2000 ppm A 80% 27151	20% 27138 +1000 ppm B 80% 27151	20% 27138 +2000 ppm B 80% 27151	20% 2729 +100C ppm A 80% 27151	20% 27138 +2000pp m A 80%	20% 27138 +1000pp m B, 80%	20% 27138 +2000pp m B, 95%	20% 27138 +1000pp m A, 80%
0 Weeks Filterable Insols	D 2274, mod	mg/100 ml	35.7	0.3	0.1	0	0.1	0.6	0.2	58.6
Adherent Insols	D 2274, mod	mg/100 ml	65.1	0	0	0.2	0.1	0.1	0	89.5

Time	Property	Method	Units	CL06-480	CL06-481	CL06-482	CL06-483	CL06-484	CL06-485	CL06-486	CL06-487
				20%AL-7128 in AL-27175-F	20%AL-7129 in AL-27150-F	20%AL-7137 in AL-27171-F	20%AL-7138 in AL-27151-F	20%AL-7141 in AL-27166-F	20%AL-7148 in AL-27166-F	20%AL-7152 in AL-27175-F	20%AL-7160 in AL-27176-F
	Total Insols	D 2274, mod	mg/100 ml	100.8	0.3	0.1	0.2	0.2	0.5	0.2	148.1
4 Weeks	Rancimat IP		h	9.35	4.63	6.98	>12	>12	4.47	6.63	5.95
	Filterable Insols	D 4625, mod	mg/100 ml	0.8	0.5	0.3	0.3	0.1	0.2	0.1	0.1
	Adherent Insols	D 4625, mod	mg/100 ml	0.4	0.3	0.4	0.3	4.1	0.1	0.1	0
	Total Insols	D 4625, mod	mg/100 ml	1.2	0.8	0.7	0.6	4.2	0.3	0.2	0.1
8 Weeks	Rancimat IP		h	4.75	3.25	5.46	>12	8.27	4.34	5.84	6.29
	Total Acid No.	D 664	mg KOH/g	0.94	1.01	1.02	1.02				
	Filterable Insols	D 4625, mod	mg/100 ml	0.5	1	0.4	0.2	0.2	0.4	0.3	1.3
	Adherent Insols	D 4625, mod	mg/100 ml	0.8	1.4	0.3	0.3	1.1	0.7	0.4	0.3
	Total Insols	D 4625, mod	mg/100 ml	1.3	2.4	0.7	-0.1	1.3	1.1	0.7	1.6
12 Weeks	Rancimat IP		h	7.78	3.43	5.86	>12	8.65	4	7	2.55
	Total Acid No.	D 664	mg KOH/g	0.94	1.01	1.02	1.02				
	Filterable Insols	D 4625, mod	mg/100 ml	0.2	0.2	0.2	0.1	0.5	0.8	0.9	1.1
	Adherent Insols	D 4625, mod	mg/100 ml	1.1	0.7	1.2	0.6	0.6	0.9	0.3	0.4
	Total Insols	D 4625, mod	mg/100 ml	1.3	0.9	1.4	0.7	1.1	1.7	1.2	1.5
	Rancimat IP		h	6.66	3.47	5.81	>12				
	Total Acid No.	D 664	mg KOH/g	0.06	0.08	0.11	0.11				

Appendix G. (Continued).

			CL06-480	CL06-481	CL06-482	CL06-483	CL06-484	CL06-485	CL06-486	CL06-487
			20%AL-7128 in AL-27175-F	20%AL-7129 in AL-27150-F	20%AL-7137 in AL-27171-F	20%AL-7138 in AL-27151-F	20%AL-7141 in AL-27166-F	20%AL-7148 in AL-27166-F	20%AL-7152 in AL-27175-F	20%AL-7160 in AL-27176-F
1 Weeks	Filterable Insols	D 4625, mod	mg/100 ml	0.6	1.1	0.8	0.5	2.2	6.4	2.3	10.9
80C	Adherent Insols	D 4625, mod	mg/100 ml	0.1	0.4	0.1	0.3	-0.3	-0.6	0.0	0.1
	Total Insols	D 4625, mod	mg/100 ml	0.7	1.5	0.9	0.8	1.9	5.8	2.3	11.0
	Thermal Stability	D 6468	% Reflectance	99.3	99.5	97.1	99	97.6	91.6	97.4	91.4
			Heptane Rinse mg/150ml	1.05	0.45	0.45	0.6	0.15	4.8	0.15	4.05

Additional Antioxidant Treated B20 Blends				CL06-547	CL06-549	CL06-551	CL06-553
				20% 27138 + 1000 ppm A 80% 27175	20% 27138 + 100 0 ppm A 80 % 27151	20% 27138 + 1000 ppm B 80% 27151	20% 27138 + 200 0 ppm B 80 % 27175
1 Weeks	Filterable Insols	D 4625, mod	mg/100 ml	4.3	15.5	0.7	11.9
80C	Adherent Insols	D 4625, mod	mg/100 ml	0.7	0.6	0.1	0.2
	Total Insols	D 462 5, mod	mg/100 ml	5.0	16.1	0.8	12.1
	Thermal Stability	D 6468	% Reflectance	89	97	99	96
			Heptane Rinse mg/150ml	7.8	19.8	0.15	0.15

In: Biodiesel Fuels Reexamined
Editor: Bryce A. Kohler

ISBN: 978-1-60876-140-1
© 2011 Nova Science Publishers, Inc.

Chapter 5

STRAIGHT VEGETABLE OIL AS A DIESEL FUEL? [*]

United States Department of Energy

Concerns about U.S. reliance on imported petroleum and fluctuating fuel prices have led to growing interest in using biodiesel, an alternative fuel made from vegetable oils. However, there is also interest in the direct use of vegetable oils as straight or raw vegetable oil (SVO or RVO), or of waste oils from cooking and other processes. These options are appealing because SVO and RVO can be obtained from U.S. agricultural or industrial sources without intermediate processing. However, SVO is not the same as biodiesel, and is generally not considered to be an acceptable vehicle fuel for large-scale or long-term use.

While straight vegetable oil or mixtures of SVO and diesel fuel have been used by some over the years, research has shown that SVO has technical issues that pose barriers to widespread acceptance.

PERFORMANCE OF SVO

The published engineering literature strongly indicates that the use of SVO will lead to reduced engine life. This reduced engine life is caused by the build up of carbon deposits inside the engine, as well as negative impacts of SVO on the engine lubricant. Both carbon deposits and excessive buildup of SVO in the lubricant are caused by the very high boiling point and viscosity of SVO relative to the required boiling range for diesel fuel. The carbon buildup doesn't necessarily happen quickly, but instead over a longer period. These conclusions are consistent across a significant body of technical information in multiple articles and reports.

A recent technical paper[1] reviews published data on the use of SVO in engines. Quoting from this paper:

> Compared to No. 2 diesel fuel, all of the vegetable oils are much more viscous, are much more reactive to oxygen, and have higher cloud point and pour point temperatures.

---

[*] This is an edited, reformatted and augmented version of a national laboratory of the U.S. Department of Energy's publication, dated April 2006.

Diesel engines with vegetable oils offer acceptable engine performance and emissions for short-term operation. Long-term operation results in operational and durability problems.

Some investigators have explored modifying the vehicle to preheat the SVO prior to injection into the engine. Others have examined blends of vegetable oil with conventional diesel. These techniques may mitigate the problems to some degree, but do not eliminate them entirely. Studies show that carbon build up continues over time, resulting in higher engine maintenance costs and/or shorter engine life. Figure 1 shows how the tendency to form carbon deposits increases with blending of a vegetable oil into a diesel fuel.[2]

Another issue that is particularly critical for use of neat (100%) SVO is fuel viscosity. As Figure 2 (next page) indicates, the viscosity of SVO is much higher than that of diesel fuel at normal operating temperatures.[3] This can cause premature wear of fuel pumps and injectors, and also can dramatically alter the structure of the fuel spray coming out of the injectors to increase droplet size, decrease spray angle, and increase spray penetration. All of these changes to the fuel spray will tend to increase wetting of engine internal surfaces with the fuel leading to increased tendency to form carbon deposits and dilute the lubricant.

The long-term effect of using SVO in modern diesel engines that are equipped with catalytic converters or filter traps is also a matter of concern. In general, these systems were not originally designed with SVO in mind, and can be seriously damaged or poisoned by out-of-spec or contaminated fuel.

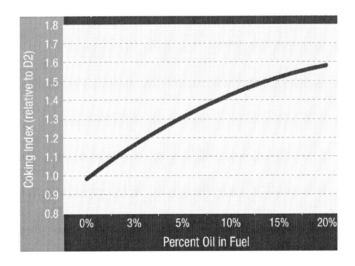

Figure 1.

BIODIESEL: FUEL MADE FROM SVO

Biodiesel is an alternative fuel that can be made from SVO in a chemical process called transesterification that involves reaction with methanol using caustic soda (sodium hydroxide) as catalyst. Biodiesel has substantially different properties than SVO, and results in better engine performance. In particular, biodiesel has a lower boiling point and viscosity than does SVO. Because of its improved qualities, vehicle and engine manufacturers are more willing to support use of biodiesel blends in their products, which will ease some of the barriers to introducing a new fuel.

The quality of biodiesel is governed by specifications developed by the American Society for Testing and Materials (ASTM). The specifications are for pure biodiesel (B100), which can be used in blends up to a maximum of 20% by volume biodiesel. ASTM specification D6751- 03a is intended to ensure the quality of biodiesel used in the United States, and any biodiesel used for blending should meet this specification. Biodiesel that meets ASTM D6751-03a is also legally registered as a fuel and fuel additive with the U.S. Environmental Protection Agency. For a complete list of ASTM biodiesel requirements, see the 2004 Biodiesel Handling and Use Guidelines at *www. nrel.gov/docs/fy05osti/36182.pdf.* In addition, the National Biodiesel Board is instituting a quality assurance program for biodiesel producers and marketers. To learn more about the BQ-9000 program, visit the National Biodiesel Board at *www.biodiesel.org.*

WHERE CAN I GET MORE INFORMATION?

- The Clean Cites activity has produced a fact sheet on biodiesel blends. It is available at *www.eere.energy.gov/ cleancities/blends/pdfs/37136.pdf.*
- The U.S. DOE's Alternative Fuels Data Center at *www.eere.energy* is a vast collection of information on alternative fuels and alternative fuel vehicles.
- The U.S. Department of Energy's Office of Energy Efficiency and Renewable Energy website at *www.eere.energy* contains information on biodiesel and ethanol production and policy.
- The National Biodiesel Board is the national trade
- association representing the biodiesel industry. Its website, *www.biodiesel.org,* serves as a clearinghouse of biodiesel related information.
- The National Renewable Energy Laboratory's Non- petroleum Based Fuels website, located at *www.nrel.gov/ vehiclesandfuels/npbf,* provides links to a variety of biofuels documents.
- The Environmental Protection Agency's Biodiesel Emissions Analysis Program at *www.epa.gov/otaq/models* contains a biodiesel emissions database.

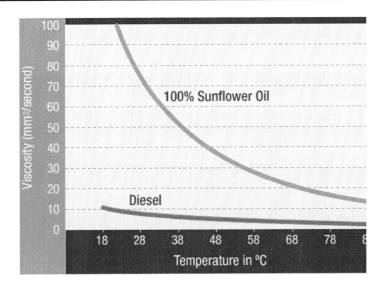

Figure 2.

End Note

[1] Babu, A.K.; Devaradjane, G. "Vegetable Oils And Their Derivatives As Fuels For CI Engines: An Overview." SAE Technical Paper No. 2003-01-0767.

[2] Jones, Samuel T.; Peterson, Charles L.; Thompson, Joseph C. Biological and Agricultural Engineering Department, University of Idaho, Moscow, Idaho, USA. "Used Vegetable Oil Fuel Blend Comparisons Using Injector Coking in a DI Diesel Engine." Presented at 2001 ASAE Annual International Meeting, Sacramento, California, USA, July 30–August 1, 2001. ASAE Paper No. 01-6051.

[3] Bruwer, J.J., et al. "Use of Sunflower Seed Oil in Diesel Engined Tractors." Proceedings of the IV International Symposium on Alcohol Fuels Technology; October 5, 1980, Guaruja, SP, Brazil.

INDEX

T

U

V

W

Y

Z